OCR PRODUCT DESIGN FOR GCSE

PHILIP CLARKE
GEOFF HANCOCK
AUSTIN STRICKLAND
BOB WHITE

HODDER
EDUCATION
AN HACHETTE UK COMPANY

Orders: please contact Bookpoint Ltd, 130 Milton Park, Abingdon, Oxon
OX14 4SB. Telephone: +44 (0)1235 827720. Fax: +44 (0)1235 400454.
Lines are open from 9.00am to 5.00pm, Monday to Saturday, with a 24-
hour message-answering service. You can also order through our website
www.hoddereducation.co.uk

If you have any comments to make about this, or any of our other titles,
please send them to educationenquiries@hodder.co.uk

British Library Cataloguing in Publication Data
A catalogue record for this title is available from the British Library

ISBN: 978 0 340 98200 6

First Edition Published 2009
Impression number   10 9 8 7 6 5 4 3 2
Year   2012 2011 2010 2009

Copyright © 2009 Austin Strickland, Bob White, Geoff Hancock,
Philip Clarke
Illustrations by Ian Foulis and Barking Dog Art

Hachette UK's policy is to use papers that are natural, renewable and
recyclable products and made from wood grown in sustainable forests.
The logging and manufacturing processes are expected to conform to the
environmental regulations of the country of origin.

Cover photo © Garry Gay/Alamy
Typeset by Fakenham Photosetting Ltd, Fakenham, Norfolk
Printed in Italy for Hodder Education, an Hachette UK Company, 338 Euston
Road, London NW1 3BH

CONTENTS

HOW TO GET THE MOST OUT OF THIS BOOK

Welcome to OCR Design and Technology for GCSE Product Design (specification numbers J305 and J045).

The book has been designed to support you throughout your GCSE course. It provides clear and precise guidance for each of the four units that make up the full course qualification, along with detailed information about the subject content of the course. It will be an extremely effective resource in helping you prepare for both controlled assessment and the examined units.

The book has been written and developed by a team of writers who have considerable specialist knowledge of the subject area and are all very experienced teachers.

The book:
- *is student focused. The aim of the book is to help you achieve the best possible results from your study of GCSE Product Design*
- *gives clear guidance of exactly what is expected of you in both controlled assessment and examined units*
- *contains examiner tips and guidance to help improve your performance in both controlled assessment and examined units*
- *provides detailed information relating to the subject content and designing*
- *is designed to help you locate information quickly*
- *is focused on the OCR specification for GCSE Product Design*
- *has relevance and value to other GCSE Product Design courses*

The book outlines the knowledge, skills and understanding required to be successful within GCSE Product Design. It is designed to give you a 'product design toolkit' which can be used to develop your knowledge and understanding during the course and support you when undertaking both controlled assessment and examined units.

Chapters 2–7 form the 'product design toolkit'. Chapters 8–11 give specific guidance about each of the units that make up the GCSE course.

▶ Unit A551 Developing and Applying Design Skills

Chapter 8 gives detailed information about the structure of the controlled assessment unit and the rules relating to the controlled assessment task you will undertake. It clearly explains what you need to do section by section and includes examiner tips to help improve your performance. Specific reference is made to the assessment criteria and an explanation is provided as to how the criteria will be applied to your product. Examples of students' work are used within the text to reinforce the requirements of each section.

▶ Unit A552 Designing and Making – The Innovation Challenge

This chapter provides detailed information relating to this unit. It gives a clear explanation of the structure of the examination and the type of task you will be required to undertake. Examiner tips and examples of students work are used to explain what you need to do in order to achieve the highest possible mark.

▶ Unit A553 Making, Testing and Marketing

Chapter 10 follows a similar format to Chapter 8. It explains the requirements of the unit section by section and includes examples of students' work and examiner tips to guide you through the controlled assessment task.

▶ Unit A554 Designing Influences

Chapter 11 is designed to help you prepare for the written examination. It clearly describes the format of the examination paper and gives examples of questions. Examiner tips are given to help you identify the type of question and the approach you should take in completing your answer.

▶ Icons used in this book

Introduction boxes provide a short overview of the topics under discussion in the section.

KEY POINTS

- Key Points boxes list key aspects of a topic.

KEY TERMS

Key Terms boxes provide definitions of the technical terms used in the section.

EXAMINER'S TIPS

Examiner's Tips boxes give tips on how to improve performance in both the Controlled Assessment and examined units.

ACTIVITY

Activity boxes suggest interesting tasks to support, enhance and extend learning opportunities.

LEARNING OUTCOMES

Learning Outcomes boxes highlight the knowledge and understanding you should have developed by the end of the section.

QUESTIONS

Questions boxes provide practice questions to test key areas of the content of the specification.

CASE STUDY

Case study boxes provide examples of how real-life businesses use the knowledge and skills discussed.

ACKNOWLEDGEMENTS

The authors would like to thank the following: Claire, Grainne and Conor White; SpeedStep – Heidi Ambrose-Brown; Nicola from Bern; Birchfield Interactive Ltd; the staff and students of St George's College of Technology, with particular thanks to John James, Sally Jones and Rob Reet; Heather Clarke; Jo Thomas; TechSoft UK Ltd; Scott Silver, Robbie McDougall and Gregory Walsh, Mill Hill School Foundation, Barnet; Shreena Patel and Lauren Ankaah, Old Palace of John Whitgift School, Croydon; Joshua Sherwood, Redbourne Upper School, Ampthill; Bonnie Pickard, The Barclay School, Stevenage; Steven Haigh, Zach Kirsch-Pinfold, Eloise Brien, Jessica Lodge, Ed Burrows, Silcoates School, Wakefield; Marc-Antoine Di Gusto, Ampleforth College; Catherine Taylor and Robert Wadsworth, Clevedon Community School; Zoe Bloss and Niazy Hazeldine, Plymouth Devonport High School for Girls; Anneka France, Chichester High School for Girls; Joanne Picketts, Cameron de Bachuus Lacey and Lauren Welfare, Ash Manor School, Surrey; Ampleforth College, York; Babington Community College, Leicester; Bolton School Boys Division; Bromfords School, Essex; Clevedon Community School, Bristol; Finchley Catholic High School; Gordano School, Bristol; Greenwood Dale School, Nottingham; Hathershaw College of Technology and Sport; Mill Hill School, London; Pimlico Academy, London; Silcoates School, Wakefield; St Annes Catholic School, Southampton; St Thomas More Language College, Chelsea; West Somerset Community College.

The authors and publishers would like to thank the following for use of photographs in this volume:

Figure 1.1 Science Photo Library/Charles D. Winters; Figure 1.2 James Dyson Foundation; Figure 1.3a Alamy/Creative Element Photos; Figure 1.3b Alamy/worldthroughthelens-décor; Figure 1.4a Alamy/Built Images; Figure 1.4b Alamy/Steve Hamblin; Figure 1.6a Fotolia.com/Andres Rodriguez; Figure 1.6b © Tijara Images – Fotolia.com; Figure 1.7a istockphoto.com /gbrundin; Figure 1.7b istockphoto.com/Tom Mc Nemar; Figure 1.8a Fotolia.com/Lorelyn Medina; Figure 1.8b istockphoto.com/Michelle Allen; Figure 1.13 Habitat UK Ltd.; Figure 1.14 Wallace & Gromit Cracking Celebration Cakes by Debbie Brown (B Dutton Publishing), available from www.squires-shop.com; Figure 1.15 "The Amazing Book of Paper Boats" Produced by Melcher Media; Figure 1.16a iStockphoto.com/Maria Toutoudaki; Figure 1.16b iStockphoto.com/Alenate; Figure 1.16c Alamy/Vario images GmbH & Co.KG; Figure 1.17 Alamy/Pictorial Press Ltd.; Figure 1.18 Fotolia.com/leafy; Figure 1.19

Corbis/Bettmann; Figure 1.20 Photolibrary/Heather Brown; Figure 1.21 copyright STIB/MIVB; Figure 1.22 Metroselskabet; Figure 1.23 Photolibrary Group Ltd/Stockbyte; Figure 2.1 iStockphoto.com/Tony Marven; Figure 2.2 © Martin Jenkinson/Alamy; Figure 2.3 Alamy/Chris Howes/Wild Places Photography; Figure 2.4 Alamy/David Perkins; Figure 2.5 Alamy/DBurke; Figure 2.6 iStockphoto.com/Leah Marshall; Figure 2.9 iStockphoto.com/Ayzek; Figure 2.11 iStockphoto.com/Katherine Moffitt; Figure 2.12 iStockphoto.com/Alex Slobodkin; Figure 2.14 iStockphoto.com/Winston Davidian; Figure 2.17 iStockphoto.com/Natalia Lukiyanova; Figure 2.19 iStockphoto.com/Marcus Clackson; Figure 2.20 Fotolia.com/Norman Chan; Figure 2.21 iStockphoto.com/David H. Lewis; Figure 3.6 Fotolia.com/Bloory; Figure 3.17 Fotolia.com/Marek Kosmal; Figure 3.21 Fotolia.com/Bunny Rabbit; Figure 4.1 Alamy/Jim West; Figure 4.4 Science Photo Library/Pasquale Sorrentino; Figure 4.9 Corbis/Moodboard; Figure 4.11 Corbis/Tony Aruzza; Figure 6.2a iStockphoto.com/Michal Koziarski; Figure 6.2b Getty Images/Stockbyte; Figure 6.2c Getty Images/Stockbyte; Figure 6.2d iStockphoto.com/Tupporn Sirichoo; Figure 6.2e Getty Images/Stockbyte; Figure 6.5a iStockphoto.com/Zak; Figure 6.5b iStockphoto.com/Wegit; Figure 6.5c iStockphoto.com/Camilla Wisbauer; Figure 6.5d iStockphoto.com/Özgür Donmaz; Figure 6.5e Getty Images/Jules Frazier/Photodisc; Figure 7.2 Rex Features/Sky Magazine; Figure 7.3 Nigel Wilkins/Alamy; Figure 7.4 iStockphoto.com/Milos Luzanin; Figure 7.5 Purestock X; Figure 7.6 Alamy/The London Art Archive; Figure 7.9a Alamy/Tom Mackie; Figure 7.9b Alamy/View Pictures Ltd.; Figure 7.10 Alamy/Geoff Wiggins; Figure 7.11 Cadbury UK Ltd.; Figure 7.12 Alamy/Jeff Morgan retail and commerce; Figure 7.15 Getty Images/Emma Lee/Life File/Photodisc; Figure 7.16 Alamy/Chris A. Crumley; Figure 7.17 Getty Images/Photodisc; Figure 7.20 Alamy/Brian Hamilton; Figure 7.21 © Galina Barskaya – Fotolia.com; Figure 7.22 © Vince Clements – Fotolia.com; Figure 7.25 © Uyen Le/bluestocking/istockphoto.com; Figure 7.26 ©Alpha Press; Figure 7.27 James Dyson Foundation; Figure 7.29 Alamy/Image Source Pink; Figure 7.30 Megaman; Figure 7.31 Science Photo Library/Simon Fraser; Figure 7.32 Everything Environmental Ltd.; Figure 7.34 Alamy/Paul Hartnett/PYMCA; Figure 7.35a Alamy/Tony Cordoza; Figure 7.35b Alamy/INTERFOTO Pressebildagentur; Figure 7.35c Alamy/Tony Cordoza; Figure 7.36a Alamy/Gary Vogelmann; Figure 7.36b Science Photo Library/Jerry Mason; Figure 7.37 Nokia; Figure 7.38 PA Photos/Sutton; Figure 7.40 Science Photo Library/Tek Image; Figure 7.41 Courtesy of Apple; Figure 7.42 Unilever UK Limited; Figure 7.43 Acme Whistles; Figure 7.44 Dr Martens; Figure 7.45 Ikea; Figure 7.46 Fotolia.com/Uhotti Figure 8.21a Photolibrary Group Ltd/Photodisc; Figure 8.21b Photolibrary Group Ltd/Photodisc; Figure 8.21c iStockphoto.com/Maurice van der Velden; Figure 8.21d Fotolia.com/Arvind Balaraman; Figure 8.21e Fotolia.com/Gary718; Figure 8.21f Fotolia.com/Jon R Peters; Figure 10.6a Getty Images/Photodisc; Figure 10.6b Photolibrary Group Ltd/Stockbyte; Figure 10.6c iStockphoto.com/Andres Balcazar; Figure 10.6d iStockphoto.com/Maksym Bondarchuk; Figure 10.6e Getty Images/Photodisc; Figure 10.9a iStockphoto.com/John Kounadeas; Figure 10.9b iStockphoto.com/MrPlumo; Figure 10.9c iStockphoto.com/Philip Barker; Figure 10.10 CLEAPSS.

Every effort has been made to trace and acknowledge ownership of copyright. The publishers will be glad to make suitable arrangements with any copyright holders whom it has not been possible to contact.

CHAPTER 1

PRODUCT DESIGN OVERVIEW

1.1 WHAT IS PRODUCT DESIGN?

'A man may die, nations may rise and fall, but an idea lives on.'

John F. Kennedy (US president, 1961–1963)

LEARNING OUTCOMES

By the end of this section, you should have developed a knowledge and understanding of:

- why product design is important to us all
- the fact that product designers work in a range of materials and improve peoples' lives.

Let's start this section off with a challenge! Can you think of anything that hasn't been designed? The chairs that we sit on, much of the food that we eat and the clothes we wear have all been carefully designed. Even a matchstick has been designed.

Fundamental to the product designer is the desire to design things that people will want and enjoy, and most important, that will improve people's lives. We come into contact with thousands of products every day. Very often, we take design so much for granted that we don't even notice it. It is only when things don't work or are uncomfortable that we begin to notice them. The door handle that is stiff, the shirt that itches, the chair that is uncomfortable; all of these things make us question why they are not working properly. In fact, things that don't work properly are often the stimulus for good product design.

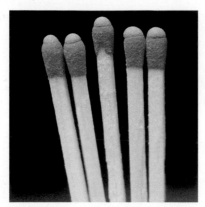

Figure 1.1 Even the humble matchstick has been designed

If you look back at the great designers in history, many of them did not work in one material alone. Arguably the greatest designer of all time was Leonardo da Vinci. Many of you will know Leonardo from his great works of art such as the *Mona Lisa* and *The Last Supper*. But Leonardo was also a scientist, a sculptor, an architect, a philosopher, an astronomer and many more things besides. True designers are not restricted by one material. They have the flexibility and creativity to work in a range of materials and draw on different technologies in order to find the best possible solutions to their problems. Some designers work more like scientists, whereas others work more like artists. For example, two of the UK's most famous chefs make well-designed products, yet work in very different ways. Heston Blumenthal is, arguably, a scientist. He uses liquid nitrogen in food preparation, and draws on scientific principles to combine very different food materials. However, he makes food products that people want to buy. Chefs such as Gordon Ramsey focus on taste and aesthetic appeal. With his thorough understanding of flavours, food materials and the art of food preparation, he produces foods that are very

appealing. The essence of design is the problem-solving process. It starts with a problem or need and is only finished when a successful product is produced. It is a creative process that draws on science and art in different measures. There is no one way to design something, but all designers follow a simple process when developing a product. The best designers take risks and are prepared to fail. Occasionally, they may seem wacky. The best motto for young designers is: 'If you always do what you always did, then you'll always get what you always got.'

Designers like Leonardo da Vinci, Gordon Ramsey, Yves St Laurent and James Dyson have all used the principle that design is a creative process that draws upon science and art in different measures.

Figure 1.2 James Dyson

▌ Principles of design

When designing a product, the designer always has to balance how it works (function of the product) with how it looks (aesthetics). Throughout the history of design, designers have always had to decide how much effort should be put into how a product looks (its form), relative to how the product works. This is known as 'form versus function'. One way towards understanding this is to think of a horizontal line. At one end of the line is the purely functional object, at the other is the purely aesthetic (decorative) product.

Every product has a function and everything has form, and a designer has to make a decision about the balance between form and function. Take a look at a wooden clothes peg. Does it need to be that shape? Does it need to be that colour? Does it need to be made from that material? A simple search on the internet will show you that pegs come in a range of colours and designs. In contrast, the function of a brooch is mainly decorative, but even that has different clasps and ways of fixing to a garment (function).

Look carefully at the two doorknobs in Figure 1.4. All they do is help a person to open or close a door. So why are they so different?

Victorian designers tended to make their designs very ornate. They believed that ornamentation (adding decoration) represented good design. On the other hand, many designers who came after the Victorians believed that ornamentation was unnecessarily ugly and that design should stem only from an object's function. So,

designers need to make decisions bearing the current fashions in mind. However, there is a lot more to design than that!

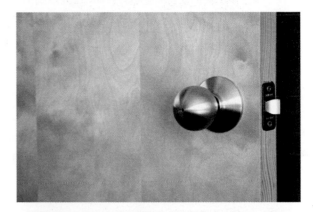

Figure 1.4 Two doorknobs

Designing is a complex activity. Not only do designers have to consider the form and the function of a product, they have to consider

Figure 1.3 Form versus function

other aspects such as fashion trends, ergonomics, the properties of materials, how much an item will cost to produce, how it will be made and cultural influences.

1.2 OCR GCSE PRODUCT DESIGN

LEARNING OUTCOMES

By the end of this section, you should have developed a knowledge and understanding of:

- the structure and content of the four units of the GCSE Product Design course
- the advantages and benefits of the Product Design course
- the fact that the four assessable units are only part of your overall studies in the subject.

You have chosen to do a very exciting and interesting GCSE within technology. It gives you a real opportunity to demonstrate what you know and what you can do in a variety of ways.

This book has been designed to really help you understand what it is you have to do in each unit to ensure that you get a good GCSE grade.

The parts of the Product Design course that will be assessed by the examination board consist of four completely different units. They can be worked on at any time during

Figure 1.5 **What the students say**

GCSE Product Design

It was that time again: year nine option choices for GCSE. I was also standing among the year nines, confused and hassled. I had all my other options planned out and I was happy with them. However, a technology choice was an obstacle in my path. At Westcliff High School for Girls, a technology – either Food, Textiles or Product Design – is compulsory and I had not chosen one yet.

Although being very hesitant at first, after hearing of the new course – Product Design – I decided to take the subject. I took it as a new challenge, maybe a risk, but it was something I had to face with a smile. I plunged into the subject with very little, if any, drawing skills and I knew that I would fall short here. I was delighted to find that the course did not involve any technical drawings, but just the normal 3D drawings that were rendered as appropriate, corresponding to their material.

Besides the course being a pilot one, the fact that all the coursework was in an electronic format using PowerPoint Presentations was also appealing. It meant that my work would be in one place and that by the end of the course I would not have the responsibility of collating piles and piles of paper (nor would I have the guilt of killing trees in the process).

Being someone who would rather work on coursework than written exams any day, I was pleased to hear that at the end of the first year my exam was not a written but a practical one. It was a six-hour Innovation Challenge that allowed me to develop my imagination, innovation and design-and-making skills. We were given a task on the day and we then had to create a product based on the specifications given. Our task involved creating something we could take to the beach, in any weather conditions. But the product had to be able to carry something. After thinking long and hard but quickly – that is what the challenge is all about – I designed a surfboard that turned into a box. It was innovative as it was not in the market and managed to achieve me an A. Despite the six hours of work, I managed to finish each of the two three-hour sessions of the first year examination with a smile.

At the start of year 11, we began our second piece of coursework – Unit 3 – where the focus was more on the making, testing and marketing side of things. The speciality of this unit (and this course) is that somewhere along the line it will involve skills you collect from other subjects. For example with Unit 3 I had to create an innovative product and I was given the freedom to choose whichever material I wanted to make my product from. I decided to take advantage of this and so I took the Food Technology root and made an innovative food product. Based on the set 'Celebrations' theme, I created an edible cake display board and edible muffin cases. Now I hope to focus on the marketing side, I have many different ways of advertising my products, one of which is to create an advertisement that can be shown on the television or even a website.

Taking Product Design has allowed me to broaden my creative side and allowed me to think, design and work with different materials confidently. It has also allowed and helped me to apply skills I already hold to make my work of a higher standard.

your course and can be examined in any order. You may think that you would have to do Unit 1, then Unit 2 and so on, but this is not the case. Some schools even do two of the units at the same time. This makes the course very flexible.

▶ Unit A551 Developing and applying design skills

This unit is all about designing. (Later in the book, Chapter 8 explains in great detail what you have to do.)

There are some very exciting aspects:

1. You can decide on the direction your design work will go. Your teachers might steer you, but using this book, you will be able to design exactly what you want once you have identified a problem to solve.

2. Remember: you do not have to 'make' what you design. This means you can come up with ideas and creative solutions that have features and functions above and beyond the solutions normally arrived at in schools and colleges.

Figure 1.7 You can design them but you don't have to make them!

3. You can consider any materials or combination of materials in your design work. You are not restricted to just wood, textiles, electronics, graphics materials or food ingredients. You can mix materials, so for example you could design a complete

Figure 1.6 Find problems to solve that interest you

in-flight meal, including the containers, the packaging and the food contents, if you wanted to.

Figure 1.8 You can choose from a variety of materials

▶ Unit A552 The designing and making innovation challenge

This examination will involve you in a designing and modelling activity. Your school arranges when you will take the exam and it takes place in a design room or a workshop rather than in the examination hall. Everyone who has completed it so far has really enjoyed it.

(Chapter 9 tells you in detail everything you need to know to do well in this exam.)

What happens?
You sit in a small group of other students, normally four of you. You sit like this even though it is an exam and, of course, you don't communicate with each other for most of the examination. But, part way through the innovation challenge you explain to the others in your group what you have been designing and they in turn explain their ideas to you. You then feed back your thoughts and ideas to the others in your group to help move their design work forward.

Figure 1.9 A group of students working on the challenge

An individual workbook is provided for you, and your making and modelling are photographed at various points and the photographs are stuck into your workbook.

Scripted, designed, modelled and photographed
Your teacher will read you instructions from a script. The same script is used for everybody in the country, and you will solve a problem and model your solution as you go. You are encouraged to be as creative as you possibly can be in solving the problem. You must try to solve the problem that you are given at the start of the challenge.

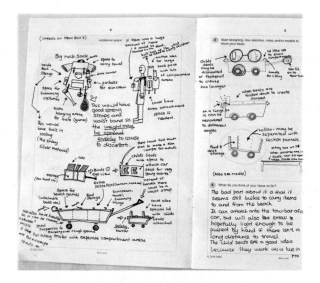

Figure 1.11

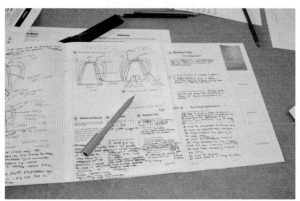

Figure 1.10 Student workbooks

Figure 1.12

You will sketch and write about your ideas as you try to solve the problem, working in different boxes in your workbook when prompted by your teacher.

Modelling your designs

You are provided with modelling materials to model your prototype and can even bring additional materials into the second three-hour session of the exam. You must not have worked on these materials at home or changed them in any way.

Photographs

At set times during the challenge, photographs will be taken of your modelling. These will be stuck in your workbook so the examiner can see how good your practical skills are at modelling and making. You can have additional photos taken of your modelling and making, which you can stick in the workbook to show your product and making in more detail. Remember to annotate these photographs.

Annotation

A **label** just simply identifies something, for example: lid, textured surface, coloured red, handle. A label offers no extra information.

A **note** gives just a little bit more information than a label, for example: lid to keep in the contents; textured surface for a good grip; coloured red so it stands out; a handle to pick it up with.

Annotation is what the examiners are really looking for, for example:

- Lid to keep in the contents so they do not spill and are kept fresh. The lid could unscrew or be a 'click type' fastening but this might not give as good a seal.

- Textured surface for a good grip – this will avoid the food mixer slipping in the user's hand. The texture could be applied to the mould and so every moulding will automatically have the same texture (and grip). This is a low-cost option because once the mould is made, thousands of food mixers can be made all with exactly the same feature.

- Coloured red so it stands out easily. Red is associated with danger and 'stop', such as in traffic lights or 'stop' buttons on domestic equipment. This means that the majority of people will understand it, so it will not need any other instruction as to its purpose.

- A handle to pick it up with that is ergonomically designed for the user. The pet-carrier is actually going to be quite lightweight, but it will be carried for long distances and also has an awkward shape, so a handle that is easy to hold and grip is very important.

Annotations are the sorts of things you would be thinking of when you are doing your designing. You should use annotation in all four units.

▶ Unit A553 Making, testing and marketing

This unit is all about *making* a quality product. Chapter 10 explains in great detail what it is you have to do but, as with the other three units, there are some exciting differences from other Design and Technology GCSEs.

The making – your 1, 2 and 3!

You can think of this unit as a sequence of activities numbered 1, 2 and 3.

The important feature of this unit is that you can make something without having to design it.

Number 1

Look at Figure 1.13. It shows a pattern that you can buy to make an Elvis Presley fancy-dress costume. It has an image of what the costume will look like on the front and details of the materials needed to make it on the

Figure 1.13 Elvis costume pattern

back. You would manufacture a costume based on the design in the pattern.

An alternative Number 1

Books, such as those in Figures 1.14 and 1.15 below, will have lots of things that you can make.

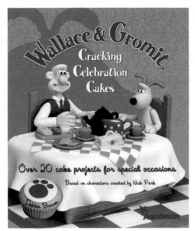

Figure 1.14 Looking for a Unit 3 starting point

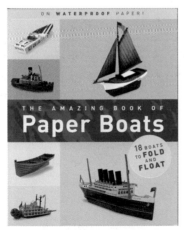

Figure 1.15 Looking for a Unit 3 starting point

You could change the designs if you wanted to, so that they suited you better. Of course, what you make doesn't have to come from these books – the designs could come from anywhere. You might find a download on the internet, or find something in another book or a pattern from a shop. If you chose to, you could make what you designed in Unit A551 or even in Unit A552. There is no restriction on what you make or the materials you use. Remember, *you* don't have to have designed it.

ACTIVITY 1

Make a list of the advantages of not having to have designed what you make in Unit A553. Think about this. In Year 8 or Year 9 you may have made something in school that was good, but not as good as you would have liked. Perhaps you didn't have as many skills then as you do now, or perhaps you only had a very short time to complete it. If you like, you could redo it and make a really good job of it for this unit!

KEY POINT

- You do not have to design it to make it.

Number 2

You do not have to restrict yourself to one material, for example wood, plastic, card, textile, ceramic or food ingredients. If you use different materials you can demonstrate more of your making and finishing skills. It is much more interesting to be working with different materials. When joining different materials together there are new challenges for you. You can once again be inventive and show off your skills to the examiner and gain valuable marks.

EXAMINER'S TIP

What the examiners are looking for is that you demonstrate a range of making skills and produce a really high quality product.

KEY POINT

• There are no restrictions on materials.

Number 3

You need to produce a production log rather than a production plan. For example, in the case of the Elvis Presley costume, there would be full instructions of how to make it in the pattern you buy. All you need to do is produce a detailed log of the stages in the making process.

KEY POINT

• There is no production planning, but a production log instead.

Avoid selecting items to make that involve very limited skills. For example, making a skateboard might involve some sawing and drilling, and some finishing off with paint or varnish. Making a scarf involves limited skills too, as does making a birthday card. Making a sponge cake only requires some weighing, mixing and cooking. Soldering a buzzer to some wires and a sensor is not overly challenging, either.

EXAMINER'S TIP

The examiners are looking at your making skills in this unit, and at your design skills in Unit A551. Be careful about what you choose to make. It must be sufficiently challenging, but you also need to consider the time you have available and whether you have the skills required to make it.

Meeting the challenge

You need to choose your making activity carefully to suit your own making skills and your school's or college's resources in order to gain the highest possible marks. Taking the simple examples given earlier, we can develop them further to make them suitably challenging.

The skateboard

Making a basic skateboard might involve some sawing and drilling and some finishing off with paint or varnish. But you could 'up skill' the project by chamfering the edges, using a lamination process to form a bend and a curve in the board, and by producing a non-slip surface for the board. Masking and painting a motif or logo will also add another skill, as would casting the wheels' housing in aluminium and boring out a hole for an axle shaft.

The card

Making a birthday card could involve only very limited skills. You could make it more challenging by embossing part of the design or adding a pop-up mechanism. If you used a scalpel to cut delicate holes through the card, you could then mount coloured materials behind the holes. If the holes were windows of a house and the coloured materials were fabric, you could make a curtain runner from some wire and beads. The possibilities to extend this type of project are endless. What could the embossing be used for and what would the pop-up mechanism do? If the house door opens, then could some sound come out?

The cake

One way of adding skills to this task would be to consider a theme such as the Mad Hatter's Tea Party. You could make a range of miniature sponge cakes with different decorations on them (showing decoration and making skills). These could be presented with the other items needed for a tea party such as cups and saucers which could all be edible too!

KEY POINT

- You do not have to plan any of your making, but you must show the important stages in your making, using photographs, with your annotations on each of them. This is called a Production Log.

Figure 1.16 Let your imagination go

Testing and promoting your product

Once you have made your product, you will test it against a specification, suggest improvements to it and explain how it (or some parts of it) would be made in the 'real world'.

Finally, you will produce a 'sales pitch' for your product, describing and giving reasons for your choice and method of promotion. You will then produce your 'sales pitch'.

KEY POINT

- Your method of promotion could be a magazine, radio or television advert, a leaflet, a point of sale display, a hoarding or any other method you think would be a good way of promoting your product.

Figure 1.18 How did Sir Alec Issigonis have such a huge influence on the motor industry?

▌ Unit A554 Design influences

Chapter 7 explains what the design influences are and how they affect the work of a designer. You will learn about them during the course and use your knowledge and understanding of them to answer questions about a variety of different products in the written examination for this unit.

You will also use the design influences when you are designing, in Unit A551.

Iconic products and trend-setting designers

As you go through your Product Design course you will begin to see and understand how important the history of product design is to us now in our everyday lives.

Figure 1.19 Why was this man so important?

Figure 1.17 Why did Mary Quant's designs and her miniskirt have such an impact on society?

Figure 1.20 How has vegetarianism had an impact on all of our lives and our eating habits?

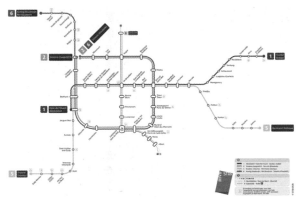

Figure 1.21 STIB map (Brussels)

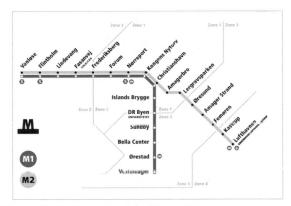

Figure 1.22 Copenhagen Metro Map

Figure 1.23 Laptop computer

Why does Figure 1.21 look so much like Figure 1.22? How does the screen of Figure 1.23 work?

You will look at the work of famous designers from the past and develop your understanding of how they have influenced your life and that of others. At the same time, you will look at some of the things they designed, for example the London Underground map, the miniskirt or the Mini Cooper, and explain why they were important.

KEY POINT

- In your exam, you will be asked questions based on the trendsetters and iconic products published by OCR for that particular year. Your teachers may work with you looking at other designers and trendsetters as well. This is a great opportunity to learn and understand more about design and design history.

The Product Design Toolbox

To be successful in your exam, you will need to be able to communicate using manual techniques and also ICT, together with Computer Aided Design (CAD) and possibly Computer Aided Manufacture (CAM). You will need to be able to annotate work correctly and also evaluate your own work and the work of others. Being able to model your thinking, research and analyse appropriately and capture your findings and observations are all skills you will also need to develop and use.

However, above all you will need to develop the skill of considering other people and situations, together with an understanding of what we call the 'real world', i.e. how people actually use products.

Chapters 2–7 form the 'Product Design Toolbox'. These chapters will equip you with the knowledge and skills required to successfully complete each of the four units that make up the full GCSE qualification.

CHAPTER 2

PRODUCT DESIGN TOOLBOX – IDENTIFYING PROBLEMS, RESEARCHING NEEDS AND WRITING A SPECIFICATION

2.1 DEVELOPING AND WRITING A DESIGN BRIEF

LEARNING OUTCOMES

By the end of this section, you should have developed a knowledge and understanding of:

- identifying a design need
- developing and writing a design brief
- identifying users and user groups
- identifying user needs and requirements.

When you hear people saying things like: 'The problem the poor receptionist had ...', or 'The postman was late this morning, I wonder why?' then they are considering other people. That is what product design is all about – trying to solve problems for others.

Figure 2.1 'The problem the poor receptionist had ...'

Figure 2.2 'The postman was late this morning, I wonder why?'

In order to solve problems through product design, you will need to develop a number of design-related skills.

▶ Developing and writing a design brief

A design brief is basically a short (two or three sentences) statement or explanation of what it is you are going to try to do. It is as simple as that.

To write a good design brief you will need to have individually identified a *specific problem or situation* that relates to a *specific user* or *user group*.

Identifying a design problem

Your *starting point* for developing and writing a design brief is to identify a problem (sometimes called the 'need or situation') and this is perhaps the most important part of the design activity. If you choose the wrong problem to solve you could severely limit your chances of success.

If you identify a problem that is too complex, then you will not be able to approach it in sufficient depth to solve it in the time you have available. If the design problem is too simple, you will restrict how many marks you can gain because of the limited opportunity to display your design skills.

Some students make the mistake of not presenting a design problem to solve, and simply state that they are going to 'design a . . .'. This is not a good idea.

- Poor examples of complex design problems would be: 'Design a submarine' or a 'Design a football stadium.' It takes years and a team of designers to design and develop these things. For the

designing unit of the Product Design course you have around 20 hours!

- Having pre-conceived ideas about what you are going to do before you even start is also problematic. An example of a poor simple design problem would be: 'Design a wooden CD rack to hold ten CDs. It will be 100 mm long by 100 mm high.' This really limits the opportunity for design because it is too detailed. There isn't much you can actually design, is there?

Anyway, who says CDs need to be stored in a rack? There are lots more exciting ways to store them than in a wooden rack.

- A good example of a design problem would be 'Storing jewellery is a problem for teenage fashion-conscious girls.' There is no immediate solution to the problem. There are hundreds of possible solutions we could explore and develop.

 EXAMINER'S TIP

Always start with a design problem and not a preconceived design idea. If you start with 'I am going to design a . . .' you will limit your ability to demonstrate your design skills and your marks are likely to be lower than they could have been.

How do you identify a design problem?
We identify design problems by looking at situations in depth.

Figure 2.3 A construction site

Figure 2.4 The simplest of situations ...

ACTIVITY 1

Look carefully at Figure 2.3 and ask yourself the following questions:

- How could you stop things falling off the scaffolding?
- How could you keep things dry when it rains?
- How can you secure things to a ladder when the workers are climbing up it?
- How could you safely dispose of rubbish from the top of the scaffolding?
- How could you stop people bumping into the scaffolding?
- How could you stop unauthorised people climbing the scaffolding?
- How could you advertise the name of the building firm on the scaffolding?

The list of questions is almost endless and, guess what, these are real problems that need solving.

ACTIVITY 2

(a) Ask as many questions about the picture (the situation) in Figure 2.4 as you can. You will then have a list of real problems that could be solved.

(b) Do the same thing with Figure 2.5.

Figure 2.5 Even going shopping ...

The user

Not identifying a suitable user, users or a user group for your design is a big mistake. You need a user to focus your designing. The user or user group is a specific individual or a specific group of people affected by a particular problem or situation. Look at the examples given below to see what you need to do to identify a user or user group:

- A poor example of a user group would be 'teenagers'. This is a huge group, made up of individual people with so many different and diverse likes, dislikes, needs, backgrounds, cultures and other factors that it would be impossible to 'group' them together using any really useful common factors.

- A good example would be 'teenage, female ice skaters'. They are still 'teenagers' with all their natural differences, but this user group has two important and easily identifiable things in common. First, they are female and second, they are ice skaters.

- A better and more tangible user group would be '16 to 18-year-old female ice skaters who use the Queens Ice Skating rink in London'. We now have a specific user group identified.

EXAMINER'S TIP

It is so much easier to gain marks if you are solving a problem for somebody else. If you only consider your own 'needs' your thinking will be narrow and will almost certainly automatically limit the marks you will be able to gain.

Figure 2.6 shows one of our '16 to 18-year-old female ice skaters who use the Queens Ice Skating rink in London'. This is regarded as evidence of the user – it is not much but it is a start towards gaining those valuable marks.

Figure 2.6 Young skater

ACTIVITY 3

(a) Make a list of the possible design problems that there might be for young female skaters that are *directly* related to their skating. Here are some examples to start you off:
- communication during the skating activities
- protecting their knees if they should fall
- keeping their skate blades in good condition.

(b) Make a second list of possible design problems that are *indirectly* linked with their ice skating. This list has also been started for you:
- club membership
- refreshments while skating
- storage of jewellery and other valuables while skating.

You may have thought of others. There are dozens of possible problems or needs.

Writing the design brief

Remembering that your design brief needs to be two or three sentences at most, let's take an example from what we have just covered and develop the associated design brief.

We have identified a jewellery storage problem for the group, which we could solve. The design brief would be:

'I am going to design a method of storing jewellery and other small valuables for 16 to 18-year-old female ice skaters. The design could be customised for different ice-rink users throughout the country.'

We have identified the problem and the users, and then opened it out a little. It makes it more challenging to say that the method needs to be used by skaters at ice rinks all over the country, rather than just at the Queens Rink in London. This way, there is lots of scope to undertake the design activity, and have some fun while gaining those valuable marks. Good grades – here we come!

KEY POINT

- You do not have to 'open out' your problem as shown in this example, but you must have a *specific user* to be successful.

ACTIVITY 4

Make a list of several possible problems or situations and identify a specific user for each of them. The more you can think of, the easier it will become!

EXAMINER'S TIP

Your portfolio should start with you providing specific details of one design problem and an associated user or user group. These should be supported by evidence and they should be summarised in your design brief.

Summary

You will need to work in the following order:

1. Identify a design problem.
2. Identify a user or user group.
3. Write the design brief.

Hannah, my older sister who is currently revising for her A-Levels, boards at xxxxxxxxxxxxx and always complains about the lighting in her room (it is either too dim or too bright), where she works, sleeps and relaxes. She recently was given a wrist watch designed by Michael Graves; a very famous and influential designer. Therefore, I have decided to base my designing around his work because Hannah enjoys his designs and I would like the user to be teenage girls like Hannah.

As this is a vital year in Hannah's school life it is essential she develops the right work technique and atmosphere in her room. She admitted that she finds it easier to work in a certain brightness of light. If Hannah is to use her main light as well as a lamp, whilst she is working, it needs to fit on her desk and create an intensity of light which she can concentrate in. However, if the lamp is desired to be used for mood lighting (e.g. going to sleep or chilling out with friends), it mustn't be too bright when the main light is switched off.

I have found that the lamp needs to be cost efficient, as students don't have much money to spare. It also must be small and neat enough to fit in a small room, with, in general, a lot of clutter. However the lamp must also be noticeable, to attract the attention of others and create a fun and relaxing mood to which the user can work or chill to.

Two of Michael Graves' lamps which I found on the internet and the watch he designed similar to Hannah's.

"...She recently was given a wrist watch designed by Michael Graves..."

Figure 2.7 An example of a student's situation and user

ACTIVITY 5

Write a design brief based on the user's needs detailed in Figure 2.7.

QUESTION 1

What do you think is missing from the example given in Figure 2.7?

In reality, identifying the user and the problem might happen at the same time, or it might be that a user leads you to a design problem. To gain maximum marks you have to be able to produce your own '1, 2, 3' list and support it with some 'real' evidence.

KEY POINT

- Your school or college may give you a theme to work to. This is sometimes called a teacher-derived or teacher-generated problem. It could be something like 'Mood Lighting', 'Charity' or 'Food on the Move'. This is perfectly acceptable and still allows you to personalise your work. Within the given context you can identify a specific design problem and user. This will allow you to undertake similar activities to those we have looked at.

Figure 2.8 An excellent example of a situation, user and brief

2.2 RESEARCH AND ANALYSIS

By the end of this section you should have developed a knowledge and understanding of:

- the purpose of carrying out research and investigation
- the techniques used to establish user needs and requirements
- how to identify relevant data
- the purpose of analysis of existing products
- techniques used to analyse existing products
- drawing conclusions from and summarising research.

As part of the design process, you will need to undertake different research and analysis activities. The purpose of these two activities is to find information that will help you meet the needs of your target users when undertaking design work. It is important that your research activities address all the aspects of your design problem. You will need to look at the results of your research, then summarise and draw conclusions from them. You should then use these results to develop points within your design specification.

Planning your research

It is important to thoroughly plan the research you intend to undertake. You may need to order books, make appointments, or wait for replies to enquiries. Thorough planning will ensure that you make the best use of your time and that you gather all the information you need in order to make your design a success.

The start point for planning your research should be to produce a research plan. Use the following questions to structure your plan:

What do I need to find information about?

- Existing products
- Size data
- Ergonomic data
- Anthropometric data
- User's opinions
- Expert opinions

Where will I find this information?

How will this information help me when designing?

Once you have established your research plan you can use the 'Five Ws' approach to develop a series of questions to explore each of the areas in depth. The 'Five Ws' are: What . . . ? When . . . ? Where . . . ? Who . . . ? and Why . . . ?

In addition to these 'Five Ws', you should ask a sixth 'W' – How . . . ? For example, How many . . . ? How often . . . ? How much . . . ?

Figure 2.9 Careful planning will lead to success

Figure 2.10 Research plan

KEY POINT

- You will definitely need a clear plan to help you **but** it will not be part of your portfolio and will not gain marks.

Primary or secondary research?

The research you undertake will fall into one of two categories. These are:

- primary research – the information you gain will be collected directly by you. This may be through interviewing experts linked to your design problem, disassembly of existing products, or through the use of questionnaires.

- secondary research – the information you gain will be based on published materials such as books, magazines or the internet. With this method of research you need to interpret data or images.

Primary research methods will usually give you more detailed information to help with your subsequent design activity.

EXAMINER'S TIPS

Be selective about what you include in the research section of your portfolio. The examiner will be looking for 'relevant material'. Plan your research thoroughly and work to your plan. Look through the information you have gathered. Only include information if it is truly relevant.

Don't be tempted to 'make things up!' The examiner will see it and you will gain no marks at all for it.

Research methods

The type of research activity you undertake will depend on the type of product you are designing. You should aim to obtain a balance of factual information and people's opinions. You should use primary research methods wherever possible. Below are the types of research methods to consider.

User research:

- interaction with target users to establish their opinions about the product you are designing. What would they want from the product? Would they require any special functions or features? This may be done by interview or questionnaire.

- observation of the target user group. What are their interests? What other products do they own? Where do they go? Are they influenced by trends and fashion?

Market research:

- product analysis of existing products to establish what similar products are available and how they meet the needs of the people using them

- interviews with users of products to establish opinions about the product, its features and performance.

Figure 2.11 **Library research**

Publication research:

- collecting information from books, magazines, newspapers and journals.

Expert opinion research:

- interviews with experts who are linked to the design problem to establish their opinions.

Product environment research:

- where the product will be used will have a significant effect upon its design. You should consider questions such as:
  - Are there any size restrictions?
  - Are there any special regulations which may affect the design of the product?
  - What are the main features of the area? Clean and dry, wet with high humidity, large amounts of dust and noise, dark or light, etc.

Internet research:

- using search engines, product reviews, retailers' and manufacturers' sites to establish information about existing products and user groups.

KEY POINT

- It is really important that you interpret the data or images you get from the internet (secondary research) and come to conclusions about the data you discover. You will gain no credit for just including downloads from the internet! It is what you do with the information that counts.

Existing product analysis

One of the most valuable sources of information for designing will come from carrying out existing product analysis. Whenever possible, this activity should be a 'primary' research activity – you should be able to physically handle and disassemble the product. Using secondary research methods at this stage will severely limit your ability to truly examine the product and the methods used in its manufacture.

You should aim to analyse at least two existing products 'in depth'. Use annotated photographs and drawings of the product or parts of the product to explain what you have found out. You should aim to analyse the

Figure 2.12 Internet research

Figure 2.13 Existing product analysis

products in as much detail as possible. Keep focused on your own design problem as you undertake the research. Remember the question, 'How is this going to help me with my design work?' as you are working.

The process:

- Start by analysing the external appearance of the product. How is the product presented? Does it have any special methods of protection for either the contents or the materials used in its construction? Is it obvious what the product is and how it functions? What colours are used? Why?

- What is the function of the product? How does the product achieve this function? Is the product successful?

- What is the target market for the product and who are the target users? Where is the product used? Where is the product sold? How is the product taken to market or the place where it is used? How many of the products are manufactured? What type of manufacturing is used?

- What materials are used in the product? Why have these materials been chosen? Could any alternative materials have been used? Could the product be recycled? How is the product assembled? What production processes were used during manufacture? What are the main stages in the production process?

Remember to draw conclusions that will help you to formulate your design specification.

EXAMINER'S TIP

Don't just label or describe features, but actually analyse why the designer has chosen them. A label pointing to a feature is worthy of little credit. Analysis linked to the feature, with reasons why it is included, will be rewarded more highly.

Questionnaires

Questionnaires offer an excellent method of establishing users' needs and opinions related to products. However, you need to consider the format of the questionnaire carefully to ensure that the information you gain is both useful and relevant.

You should plan your questions so that you only obtain 'closed' responses. Closed questions allow 'Yes' or 'No' answers or have 'tick the box' responses. This type of question is much easier to analyse.

You should avoid 'open' response questions. These allow people to express their views freely and are very difficult to analyse. However, open questions may be suitable for use in a 'one-to-one' interview with an expert.

Figure 2.14 Questionnaire research

Think about the product you are designing and the information you need to find. Structure the questions to ensure you get the information you need.

EXAMINER'S TIP

When you have completed the planning of your questionnaire, examine it and justify why each question is included and what purpose it has. If you cannot explain why it is included, it should probably be removed. Test your questionnaire with someone to see if they understand it fully and are able to respond in the way you anticipated.

KEY POINT

- All of your questions need to give you 'meaningful' responses. This will give you some real data to inform your design work. Asking what someone's favourite colour is, or what age they are, is often a complete waste of time because those facts probably have no bearing on the design problem in question.

Show the results of your questionnaire graphically. Software programs such as Excel® will allow you to quickly produce graphs of your results. Keep the graphs small to ensure you use space in your design portfolio effectively.

Look at your results and draw conclusions. Remember that these conclusions should then form points in your design specification and you should be able to justify them.

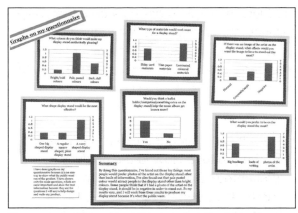

Figure 2.15 Questionnaire results displayed graphically

Size data

Depending upon the type of product you are designing you will need to research and record data linked to sizes. This may relate to the user of the product or the items that your product will contain, store or hold. General research linked to anthropometrics and ergonomics is worthy of little credit. Statements explaining what ergonomics and anthropometrics are will actually gain no credit at all.

Anthropometrics is the measurement of the human body. It provides information relating to the specific physical dimensions of humans and can be targeted to specific age or gender groups. An example might be the wrist sizes of teenage girls.

Ergonomics relates to the use of anthropometric data to ensure that items that are designed are suitable for use by people in the intended target market. Examples of the use of ergonomic data would be the consideration of the size of buttons on a mobile phone keypad, or the consideration of head size for the design of a child's cycle helmet.

Your size data should include details of the sizes of any products or items that will have an impact upon the design of your product, e.g. it would be impossible to design a DVD and games storage system without knowing how big the DVDs and games are.

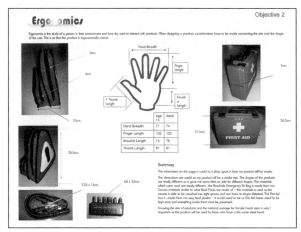

Figure 2.16 Student's ergonomic research – roadside emergency kit

Remember, all data that you include in your design portfolio must be relevant to your product.

Environmental, social, moral and cultural Issues

As designers of products, we need to be aware of the world around us and the issues people face who live in different areas of the world. Within your research, it is important to give consideration to these issues and how your product will affect them. You will find that many of the issues are interlinked with each other.

Environmental issues are becoming increasingly important. Research is clearly showing the effect that we humans are having on our planet. The use of finite resources to manufacture materials and products needs consideration by all designers

of products. Within your research, you should give consideration to how you can control the impact of your product upon the environment. This could be achieved by:

- designing products to have a long product life
- manufacturing products using recycled or recyclable materials
- planning the use of materials to avoid wastage
- designing products that have a second 'end of life' function
- using materials that are sustainable
- reducing the use of packaging
- designing products that are energy efficient.

KEY POINT

- You will gain no credit for just saying 'My design will have a long product life.' You will have to give details and explain what you actually mean by the comment when related to your design problem.

Social issues relate to the consideration of the product and the effect it will have upon the society where it is used. You should also consider the social messages that your product or the advertising for your product portrays.

Moral issues relate to human behaviour and what is generally regarded as being acceptable. A product that encouraged behaviour that people would find unacceptable would not be well received by the public. A product that is manufactured in a developing economy by people working in poor conditions and receiving low pay may also not be acceptable because of the associated moral questions.

Cultural issues relate to the need for the designer to understand different cultural backgrounds, beliefs, lifestyles and religions so that the products they design do not cause offence to people.

ACTIVITY 6

Explain the term 'fair trade'. How does it affect product design?

Your summary and conclusions

Having carried out the research activity it is important that you use the results of your work effectively. The findings of your research should be summarised with conclusions being drawn. When considering the conclusions to each part of your research, it would be useful to keep considering the question 'How does the information I have found help with my design work?'

The conclusions you reach should give you valuable information that should help you to develop your design specification and support your design work.

EXAMINER'S TIP

You may find it useful to produce conclusions on each part of your research work as you complete it. These conclusions should then be brought together on a single summary sheet that should be placed at the end of the research section. This will help you formulate your design specification.

2.3 YOUR DESIGN SPECIFICATION

LEARNING OUTCOMES

By the end of this section you should have developed a knowledge and understanding of:

- the importance of the design specification
- how to write a detailed and justified design specification.

You have identified a design problem, a user or user group and you have written your design brief. You have undertaken research, delving deeper into the user requirements, and you have also looked in detail at a number of existing products. You now have information and data from the analysis of your research that will inform your designing. What you need to do now is put all this information into an easy-to-use format. We call this your 'design specification'.

How important is a design specification?

This course comprises of four separate units. A design specification is required and used in each of the four units. It is a key aspect of this course and of any design activity:

- Without a good design specification it is impossible to design in a meaningful way.

- It is impossible for you to judge your success in solving the design problem or test your outcomes unless you have a good specification in place.

Figure 2.17 'My pen design is perfect – I just forgot to take the size of the user into account!'

Your design specification is your 'guidance list' of things to address when you are designing and to test against when you have finished designing or making.

 EXAMINER'S TIP

There is no set way a specification should be written, but you will find that a numbered or bullet-point list is the best way to order your specification points. This makes it very easy to refer back to the various points when you are designing or testing.

Key Points	Details
Function	The purpose of the product I am designing is to allow the brides to identify the guests that will be visiting their wedding without having to embarrass themselves in front of the guests you are talking to. It will tell the brides the personal details of the guests they have invited to their wedding.
Performance	The product I am designing can be placed in a handbag which she will be carrying with her or she will wearing on day she gets married. The electronic device will match with the dress the will be wearing on their wedding day
Size	The size of the product should be big enough for the bride to see information from it and it should be well discrete which the bride will be holding and if the guests did see it they will be quite offensive about how the bride doesn't know who the guests are
Weight	It matters a lot if my product is light or heavy. This is because that product will be held or worn by the bride all day long on her wedding day so it should be light as possible.
Target market	The group of people I am aiming my product is at the people who are getting married at the age of 20 – 40. The product I am designing creates an image of style, fashionable and wealthy
Life in service	This design is expected to last for about 2-3 years because mostly everyone who is getting married today and over the next year, forget their 3rd or 4th cousins, family or friend and the material I am going to use will last a long time and could use it afterwards to upload different information

Key Points	Details
Ergonomics	I will make sure that the electronic device is easy to hold or worn for quite a long time and make sure that it is not that heavy. I also have to make sure that it looks like a product made by a professional – not that big or not that small and it looks really stylish and cool.
Materials	I would have to make sure that the product is made out of a material that is waterproof so that the electronic device doesn't get wet and blows up and that it is quite strong, cheap (so that they can buy it really easily).
Safety	I have to make sure that there are no sharp bits on the electronic device because they could hurt their hand whilst using the electronic device.
Cost	The product should cost at a low price in order to allow the consumers to buy the product I am designing really easily and that more people buy it.
Reliability	Before my product is sold, it will be tested to find out the different problems that are held in this product.
Manufacturing quantity	This product will be one-batch product because it should match the theme of the wedding dress. The electronic device will be made in factories and it should be able to match theme of the dress that will be worn on the wedding day.
Possible conflict	This should be quite trendy, cost less and should match with the wedding dress the bride will be wearing. This is to make sure that the consumer can buy it easily and the product is attractive so that the

Figure 2.18 An example of a specification structured using 'Key Points'

 EXAMINER'S TIP

One specification point to try to avoid is 'Cost'. It is very difficult to make genuine comments about the cost in relation to the user and even more difficult to relate costs to the manufacture of a product, especially in the time you have available. Unless any specification point is meaningful it is best avoided.

What points will your specification need?

There is no set answer to this question as each design activity is different from the next. The list of key points used in the example shown in Figure 2.18 works well for this project but probably will not work for yours.

Probably the best way to start your specification is to use the information gained from the research you have carried out. If you analyse the results of your research and create a summary of your findings, you will find that the specification almost writes itself.

Let's see how we might approach the writing of a design specification – again, it is a process in three stages (1, 2, 3) for you to learn and develop.

▶ Strategy for developing specification writing

Using disassembly and product analysis – your 1, 2, 3!

KEY POINT

- This section is using product analysis to show how you can master the skills of specification writing. There is a whole section earlier in this chapter on how to undertake Product Analysis which you will need to read and then practise over and over again.

Figure 2.19 Bus shelter

Look at Figure 2.19. The term 'disassembly' means looking at a product and breaking it down into simple or basic components.

Stage 1
Looking at the bus shelter and visually disassembling it, we can easily see that it has:

1. a roof

2. a back panel

3. a side panel at one end

4. an advertising panel at the other end

5. a light inside it

6. a gap between the panels and the ground

7. an information panel on the back panel

8. a seat

and that

9. it uses a frame construction.

This is a simple list of the parts of the bus shelter – a good way to start off.

Stage 2
What we need is some detail and reasons why these parts are the way they are, so let's investigate further. The bus shelter has:

1. A *curved* roof *that protects people waiting for the bus from inclement weather.* **The curve of the roof means the rain will run off very easily and there will be no puddles of water forming above the users' heads.**

2. A *transparent* back panel *to protect users from the wind and rain.* **It also allows the user to see behind them so they are aware of anyone approaching the shelter.**

3. A *transparent* side panel at one end *to offer weather protection* **but which is set back to allow clear vision to see when the bus is coming.**

4. An *illuminated* advertising panel at the other end, *which will generate revenue for the bus operators by selling advertising space.* **It will also give the users something to look at and occupy them when they are waiting for the bus.**

5. **The side panel and advertising panels can be swapped around, so that the shelter can be used when traffic operates on the opposite side of the road. *This is important when exporting goods to other countries, as some of them drive on the left and some on the right.**

6. A light *that has two functions. First, it makes the user feel more comfortable because they are not waiting in the dark and second,* **it allows the bus driver to more easily identify the bus stop in the dark or inclement weather.**

7. A gap between the panels and the ground *that allows water to run away and avoids puddles forming inside the shelter.* **Additionally, it allows airflow to keep the shelter fresh. Another advantage is when the shelter and surrounding area are cleaned it is easier to sweep and remove any litter because the back and side panels are not an obstruction to the cleaning activity.**

8. An information panel on the back panel *to inform users about the services which operate from that bus stop.*

9. A seat *to offer comfort to users while they are waiting for a bus.* **The size and angles of the seat are such that it is impossible for somebody to lie down on it, which is unsocial and undesirable.**

10. Uses a frame construction *that allows for easier transport of the shelter to the site.* **Also, the prefabricated nature allows for changes to the components to be made (see earlier point above*) and also sections can be joined together to make a longer or shorter shelter.**

The additional information is in *italics* and makes the analysis so much more detailed and helpful to you.

The **emboldened italics** go even further and really give a good understanding of the aesthetics, components and function of the bus shelter.

 ## QUESTION 2

(a) Did you spot the *? Was it easy to refer back to the other point? Although the specification points were numbered, the * did not link directly with a 'specific' specification point and so leaves us all searching for the connection being made!

(b) Did you also spot the additional point, point number 5, about swapping the panels around? Smart stuff!

What has this got to do with writing a specification?

It's your number '3'. Well, it is actually more like a 'Ɛ' in other words, the reversal of what we have just done with the product analysis.

Doing stage 3 without 1 and 2

Gaining marks is easy when you know how. Take this example. A very basic specification for a bus shelter might look like this.

My bus shelter would need to have:

1. a roof

2. a back panel

3. a side panel at one end

4. etc.

5. etc.

All we have done is use the information we found out during our product analysis of the bus shelter at the simplest level.

Justification

All 'justification' means is giving a reason for something. Look at these examples.

- It needs to be 350 mm long *because*............

- It must have two separate compartments *so that*............

- It must have a Velcro fastening *as a precaution*............

- It will have a starch content *to allow* ....

- *So that it* revolves it will need a ....

- *To enable it to*............it will need to be heated quickly

- Tear resistance is vital *to ensure*............

Add justification to your stage 3

So, gaining even higher marks is easy when you know how. A good 'justified' specification for a bus shelter might look like this.

My bus shelter would need to have:

1. a roof to protect people waiting for the bus from inclement weather. *A curved roof would allow the rain to run off very easily and so there will be no puddles of water forming above the users' heads.*

2. a back panel, which is *transparent to protect users from the wind and rain and allows the user to see behind them so they are aware of anyone approaching*

the shelter. This is especially important for younger children and older people as they are sometimes quite easily intimidated.

And so on. This would then be a very good specification indeed and, if it were for your design work it would gain the highest marks.

KEY POINT

- There is no set number of points for a specification. It is unlikely that a specification will have enough content with much less than about eight different specification points. If you find you are going much above 14 points then you are perhaps going into too much detail.

Practice makes perfect

Look at Figures 2.20 and 2.21.

Figure 2.20 In-flight meal

Figure 2.21 Fireman's jacket

ACTIVITY 7

(a) Working in pairs, use Figure 2.20 to undertake product analysis by visually disassembling the 'in-flight meal'. Then turn your product analysis into a design specification for the in-flight meal.

(b) Now do the same with the fireman's jacket in Figure 2.21.

EXAMINER'S TIP

Compare the results with your partner as you go along. Of course, when you do it for real for your controlled assessment you will have to do it totally on your own, but doing it this way now will provide very good experience.

QUESTION 3

What do the following terms mean? Jot down a brief definition for each one.

- disassembly
- justification
- product analysis
- specification
- user
- practice.

Of course, if you are designing something, you cannot disassemble the product because you haven't designed it yet. But with enough practice it will be easy for you to write a specification because you will understand the structure and process you need to go though.

EXAMINER'S TIP

Always avoid the stating the obvious. For example, saying things like:

- 'My product will be durable.' What does that actually mean?
- 'My celebration dish will taste nice.' People's tastes differ, don't forget! What does 'taste nice' actually mean?
- 'The product will need to be light.' Does this mean light in colour or light in weight? Why does it *need* to be light? Light in relation to what? An elephant is lightweight if you compare it with a Jumbo Jet!

EXAMINER'S TIP

Something that is 'specific' means it is 'individual' and 'unique' to the situation. That is where we get the word '*specific*ation' from. So it is now obvious that saying something like 'It will adjust to any size,' 'it will suit anybody' or 'it will be available in every colour' just has no meaning or value.

So . . . you need to produce something that is 'specific' and 'detailed', as in the following example:

The product will need to be **light enough in weight** so that an **average eight-year-old child** can carry it around with **minimal effort**. My research suggests that **four kilograms** would be the **heaviest** it could possibly be.

Do you see how it works? Use 1, 2, and 3 to learn and practise, and then apply it to your own design work.

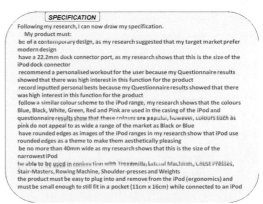

SPECIFICATION

Following my research, I can now draw my specification.
My product must:
be of a contemporary design, as my research suggested that my target market prefer modern design
have a 22.2mm dock connector port, as my research shows that this is the size of the iPod dock connector
recommend a personalised workout for the user because my Questionnaire results showed that there was high interest in this function for the product
record inputted personal bests because my Questionnaire results showed that there was high interest in this function for the product
follow a similar colour scheme to the iPod range, my research shows that the colours Blue, Black, White, Green, Red and Pink are used in the casing of the iPod and questionnaire results show that these colours are popular, however, colours such as pink do not appeal to as wide a range of the market as Black or Blue
have rounded edges as images of the iPod ranges in my research show that iPod use rounded edges as a theme to make them aesthetically pleasing
be no more than 40mm wide as my research shows that this is the size of the narrowest iPod
be able to be used in conjunction with Treadmills, Lateral Machines, Chest Presses, Stair-Masters, Rowing Machine, Shoulder-presses and Weights
the product must be easy to plug into and remove from the iPod (ergonomics) and must be small enough to still fit in a pocket (11cm x 16cm) while connected to an iPod

Figure 2.22 Specification informed by research

Figure 2.22 is an example of how good research can inform the design specification and so support the subsequent design activity.

CHAPTER 3

PRODUCT DESIGN TOOLBOX – GENERATING DESIGN IDEAS

By the end of this section, you should have developed a knowledge and understanding of:

- different approaches to designing
- drawing techniques to produce both two-dimensional and three-dimensional images
- enhancement techniques to improve communication of your design ideas
- annotation of designs.

Within the Product Design course you will undertake a number of design activities. Your teacher will use these to develop your knowledge of materials and design. You will need to be able to demonstrate your design skills in the controlled assessment units and in the external examinations. It is therefore important that you develop different techniques to communicate your design ideas effectively to other people.

3.1 GENERATING AND COMMUNICATING DESIGN IDEAS

EXAMINER'S TIP

Remember that you will not meet the examiner who will assess your work. They will not be able to question you about your design. You will need to tell the examiner the 'story' of your design, the decisions you have made and your reasons for making them, through your design folder.

The product design work you will undertake as part of this course will require you to consider the shape and form of products. In doing this work you will need to use a number of different communication methods, including drawing, sketching and photographs. These different techniques will be used to:

- generate and record design ideas

- develop and make improvements to designs
- present ideas to others
- evaluate prototype solutions and suggest further development.

◗ Getting started

You have completed your research and drawn up a design specification. Now it's time to put some ideas down on paper, but how are you going to approach this stage of the design work?

Figure 3.1 Capturing your ideas

The first thing to remember is that this stage should not be isolated from the work you have already completed. The information you have gained when researching 'user' needs and analysing existing products should be used to influence your design ideas. It is also important to remember that your design ideas need to satisfy the requirements of both your design brief and specification.

So how do you start? You may find these tips useful:

- You may already have some idea of what your product will look like. Don't limit yourself to one idea. The examiner will be looking for a 'range' of possible solutions to your design problem.
- Use your knowledge of existing solutions. Designers may create totally new products

(inventions) or examine existing products and develop ways of improving them (innovation). Try to look at existing products differently. This is called lateral thinking. For example, does a skateboard need four wheels? Do light switches need to go on the wall? Do lights need to be mounted on the ceiling? Could a USB memory stick be a cuddly toy gift? Does a dice need to be cube shaped?

- Explore possible ideas in a general way. Produce ideas quickly. Try not to limit your ideas because of perceived problems or preconceived ideas. As you develop ideas further you will be able to resolve issues with your design.
- If you are having difficulty getting started, write down all the words that you associate with the product you are designing. Select some of the words and try to use them to stimulate your ideas.
- As well as drawing your whole product, draw ideas for parts of the product and show different alternatives.
- Use annotation to explain your ideas and your opinions of them.
- As your ideas develop, you may need to use modelling materials to create prototypes to solve design problems.
- Use CAD (computer-aided design) as a 'modelling' aid to develop the design, rather than just to produce a final design drawing.
- Review your ideas against the specification and design brief. Are the ideas suitable solutions for your design problem?
- Identify and give justification for which design idea is the most suitable solution to your design problem.

Remember that there is more than one way to approach design work. Using different materials will involve you in the use of different approaches to both design and development of products. For example, the approach to modelling/prototyping ideas will be different for a food-based product and a textiles-based product.

Presenting your ideas to other people and gaining their thoughts and feedback can be very useful. Other people may identify other features to your design or could suggest a whole different idea. Try discussing your ideas with other people in your group.

SCAMPER

When developing your ideas further it may be beneficial to use the SCAMPER technique. This method helps you to think about possible improvements or developments of an idea by giving you a structure to follow:

- **S**ubstitute – consider alternate components, materials, joining methods, assembly methods.
- **C**ombine – mix parts of your ideas together to form new ideas.
- **A**dapt – think about your design. Could it function differently, could it be combined with another product? Adapt your ideas.
- **M**odify – change the shape or dimensions of part of your design or the whole design.
- **P**ut to another use – could your design have a second function? Could this function be at the 'end of life' of the original product?
- **E**liminate – simplify your design. Does it need all the functions/parts you have designed?
- **R**everse – turn your idea upside down or inside out. Look at it differently.

3.2 DRAWING AND COMMUNICATION TECHNIQUES

The drawing and communication techniques you use during the designing stage must clearly show your design and the decisions you have made. The following sections outline some of the techniques you could use while generating and developing your design ideas.

Freehand sketching

The best method to use for generating and recording design ideas is freehand sketching. Using sketching allows you to take the idea you have thought of and quickly record it on paper. Having a visual record of your alternative design ideas will allow you to compare them and identify the best idea or parts of ideas to solve your design problem.

A visual record is also important as it allows your teacher and the examiner to assess your design capability.

Freehand sketching is done without the use of any drawing instruments or aids such as rulers, set squares, or compasses. The 'flow' of your ideas is not interrupted by the need to use drawing aids. Freehand sketching can be used to show two-dimensional (2D) and three-dimensional (3D) representations of your ideas and parts of your ideas.

You can quickly show alternative methods of construction and operation together with the shape and form of your ideas. As well as sketching your ideas, add detailed notes to your design to make sure that you are getting your ideas across clearly to the people who will look at your work.

You can use a variety of drawing equipment to produce your sketches, including pencils, ballpoint pens, coloured pencils and fine-line markers.

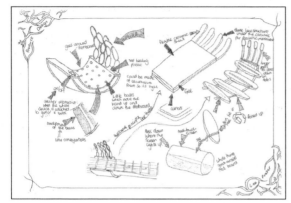

Figure 3.2 A student's design ideas for a guitar aid

When producing design sketches and formal drawings of designs, you can choose from a number of different drawing techniques, depending upon the artefact you are designing and the views of the object you want to show. The following are all useful methods of communicating your design ideas.

Oblique drawing

Oblique drawing is probably the simplest method of producing a three-dimensional (3D) representation of a design idea. Start by drawing the front view of the object and then add sloping lines at a 45-degree angle from the corners of the shape. Complete the drawing by adding lines to represent the rear of the object. These lines will represent the depth of the object you are drawing.

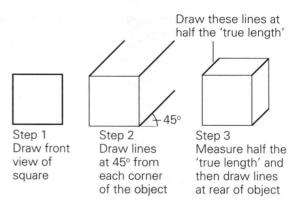

Step 1
Draw front view of square

Step 2
Draw lines at 45° from each corner of the object

Step 3
Measure half the 'true length' and then draw lines at rear of object

Figure 3.3 Constructing an oblique drawing

You will find that if you draw the object using its actual measurements it will look distorted, as shown by the drawing of the cubes in Figure 3.4. To overcome this problem you will need to scale down the length of the lines that show the depth of the object. These are normally drawn at half their true length.

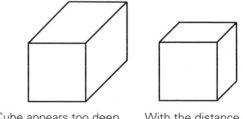

Cube appears too deep when all lines are drawn at true length

With the distance of the 45° lines halved the cube looks in proportion

Figure 3.4 Distortion of oblique drawing of cube

Isometric drawing

Isometric drawings tend to give more realistic representation of objects than oblique drawings. When drawing in isometric, you must follow two basic rules:

- All vertical lines on the object remain vertical on the drawing.

- All horizontal lines on the object are drawn at 30 degrees.

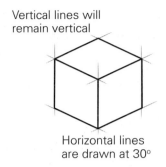

Vertical lines will remain vertical

Horizontal lines are drawn at 30°

Figure 3.5 Isometric drawing

EXAMINER'S TIPS

- In your design work, use freehand isometric and oblique sketching to help develop your designs to satisfy your design brief and specification.

- You can add colour and shading to your sketches and to your formal drawings to enhance their appearance and enhance your design communication.

Perspective drawing

Perspective drawings show objects in a similar way to how we would see them. Because of this, perspective drawings can

give the most realistic representations of designs and they are also easy to understand. Perspective drawing takes into account the fact that lines appear to converge and meet at a vanishing point. This effect can be seen if you look down a railway line, corridor or a long tunnel.

Two types of perspective drawing are used within product design – one-point perspective and two-point perspective.

One-point perspective

This is the simplest type of perspective to draw because it is based upon a flat view of the object, with all horizontal lines converging to a single vanishing point. Figure 3.7 shows the block letter 'T' drawn in single-point perspective. Initially, the block letter was drawn. The vanishing point was then marked on the page and faint lines (these are called 'construction lines') were then drawn from each corner of the letter to the vanishing point. The back edges of the letter were then drawn in. The position of the back edges has to be estimated. It is not possible to measure the depth when using perspective drawing.

Figure 3.6 Rail track appears to converge in the distance

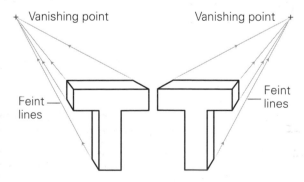

Figure 3.7 One-point perspective drawing

EXAMINER'S TIP

Being able to draw really faint construction lines is a great skill to learn and will make any of your drawing work that much more effective. As they are so faint there is no need to rub them out afterwards.

The position of the vanishing point can be varied depending on whether you wish to view the object from the left or the right, and from above or below.

Drawing curved or circular shapes in one-point perspective is relatively simple. A flat view of the object is drawn and then the depth lines are projected back to the vanishing point. This is shown in Figure 3.8.

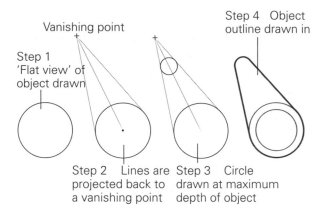

Vanishing point

Step 1 'Flat view' of object drawn

Step 2 Lines are projected back to a vanishing point

Step 3 Circle drawn at maximum depth of object

Step 4 Object outline drawn in

Figure 3.8 Drawing curved shapes using one-point perspective

ACTIVITY 1

Using one-point perspective, draw simple shapes or box letters. Draw the shape or letter with the vanishing point to the left, right, above and below the object to observe the effect this has on the drawing.

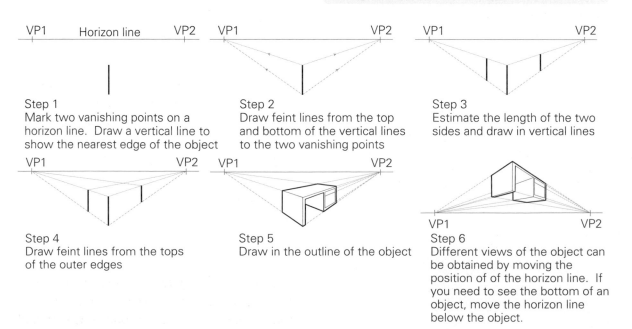

Step 1
Mark two vanishing points on a horizon line. Draw a vertical line to show the nearest edge of the object

Step 2
Draw feint lines from the top and bottom of the vertical lines to the two vanishing points

Step 3
Estimate the length of the two sides and draw in vertical lines

Step 4
Draw feint lines from the tops of the outer edges

Step 5
Draw in the outline of the object

Step 6
Different views of the object can be obtained by moving the position of of the horizon line. If you need to see the bottom of an object, move the horizon line below the object.

Figure 3.9 Two-point perspective

Two-point perspective

This method of perspective drawing will give a more realistic view of the object. The object being drawn is placed at an angle with a corner of the object appearing to be close to the person viewing the drawing. This method of perspective drawing uses two vanishing points. The horizontal lines of the object will recede in different directions, converging at the two different vanishing points. Figure 3.9 shows the method of producing a two-point perspective drawing.

Crating

Drawing box shapes is relatively easy. Unfortunately, not many of the objects you need to draw when designing are box-shaped. It is often difficult to draw complex designs that involve curves or more than one part. Crating allows you to draw simple geometric shapes and then add more detail to these shapes to create the design drawing of your object. Figure 3.10 shows design ideas for a video camera that have been drawn using the crating technique.

To use crating, first draw the crate using whichever of the 3D drawing techniques you wish. The crate needs to be drawn really feintly – remember feint construction lines. Then draw in the object, and add detail to the design.

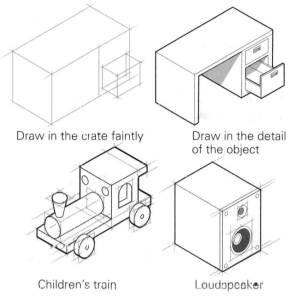

Draw in the crate faintly | Draw in the detail of the object

Children's train | Loudspeaker

Figure 3.11 Crating

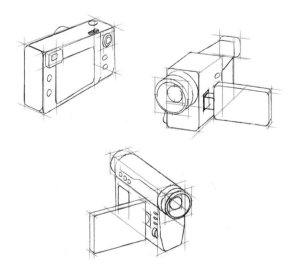

Figure 3.10 Video camera drawn using crating

ACTIVITY 2

Look around your classroom or at home for objects that you can draw. Use the crating technique to create drawings of the objects. Start with simple objects that are mainly rectangular in shape. As you feel more confident, move on to more complex objects that have curved surfaces.

Use of grid papers

If you find sketching design ideas difficult you could use grid papers to help you. You can either draw directly onto the grid or use the grid as a backing paper, where the grid is placed under the plain paper. This allows you to see the grid lines though the plain paper and then use these as a guide when drawing. Figure 3.12 shows a design drawn on isometric grid paper.

Ellipses

When a circle is viewed at an angle it becomes an oval shape, which is called an ellipse. Drawing ellipses freehand can be difficult. By following the technique shown in Figure 3.13 you will be able to draw accurate ellipses whether you are using oblique, isometric or perspective drawing.

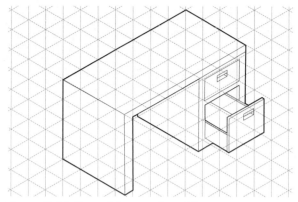

Figure 3.12 Use of isometric grid paper

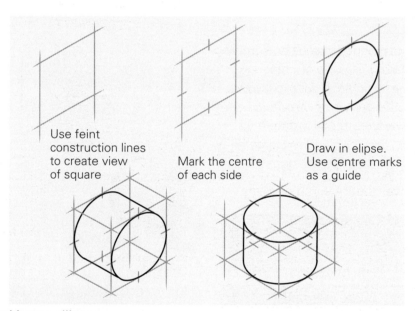

Use feint construction lines to create view of square

Mark the centre of each side

Draw in elipse. Use centre marks as a guide

Figure 3.13 Sketching an ellipse

3.3 ENHANCEMENT TECHNIQUES

All of the techniques described so far will help you to produce simple line drawings to aid the communication of your design. The lines will tell you and others what shape the product you have drawn is by showing the edges of the product. To enhance simple line drawings further, you will need to add tone and contrast to the design.

If you look at any 3D object, you will notice that some surfaces are darker or lighter than others because of the effects of the direction of light falling upon the object. Using shading to create different tones upon the object you have drawn will make the object appear more solid. Shading can also be used to indicate the types of material that the product is manufactured from.

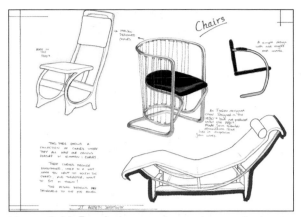

Figure 3.14 Rendering used to enhance design communication

Pencil shading

A simple method of producing tones on a drawing is to use pencil shading. The use of a medium-grade pencil such as an HB or B will allow you to create a range of different tones from dark, almost black, to very light, almost white.

Angling the pencil to allow more of the pencil lead to be in contact with the paper will allow greater control over the tone you produce.

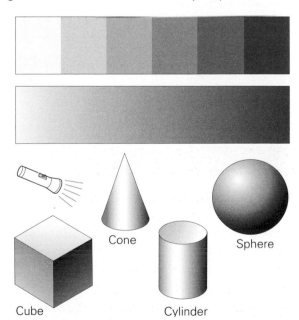

Figure 3.15 Pencil shading

ACTIVITY 3

Copy the objects from Figure 3.15 and use either an HB- or B-grade pencil to reproduce the tonal effects shown.

Dots and lines

It is possible to create tonal effects using dots and lines. The basic concept behind this technique is that lines or dots that are close together will create dark areas of tone, and lines and dots that are spaced further apart will create light areas of tone.

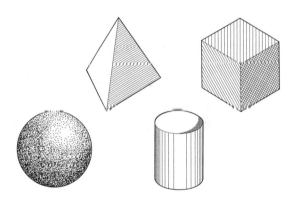

Figure 3.16 Rendering using dots and lines

Coloured pencils

Coloured pencils provide an inexpensive, quick and effective method of adding colour to designs. The hardness of the lead in coloured pencils varies between different manufacturers and it is often better not to buy the cheapest ones you can find. To create dense areas of colour, it is necessary to use crayons with softer leads. The best effects from using coloured pencils are obtained by gradually building up layers of colour until you obtain the colour you want. As with pencil shading, the pencil should be

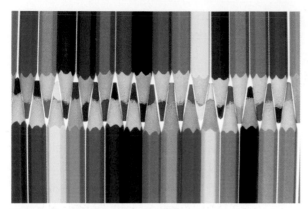

Figure 3.17 Coloured pencils

held at an angle to make sure that more of the pencil lead is in contact with the paper. To show the effects of light upon the object, it can be useful to have a lighter and darker version of the same colour crayon.

Thick and thin line technique

Using thick and thin lines on your design drawings can give a type of shadow effect that will give the object more impact. Applying this technique to your sketches and drawings is very simple. You need to follow two basic rules:

- A line where two surfaces meet and where only *one* surface can be seen is drawn as a **thick** line.
- A line where two surfaces meet and *both* surfaces can be seen is a **thin** line.

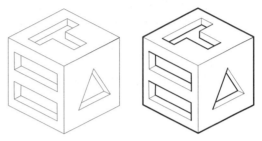

Figure 3.19 Thick and thin line technique

Marker pens

Coloured felt markers are widely available and relatively inexpensive. They are available in a wide range of colours and sizes. Two main types of marker are available:

- Water based – these are ideal for fine

Acrylic
White 'highlights' created using a rubber to create effect of shiny surface.

Aluminium
Shade silvery grey and pale 'highlights' using a rubber to give a 'satin effect.'

Plywood
Use brown, yellow and orange crayons. Draw the grain on the top surface, then draw in veneers.

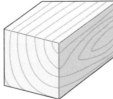

Natural wood
Use brown, yellow and orange crayons. Draw the grain and add end grain pattern. Colour surface of wood.

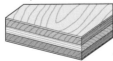

Chipboard
Use yellow and brown crayons. Colour the surface then add random dots to reflect the appearance of chipboard .

Figure 3.18 Pencil crayon rendering

detail and rendering small areas of drawings. They tend to dry very quickly which can give the drawing a 'streaky' appearance where lines overlap each other.

- Spirit based – these are more expensive than water-based markers, but are more suitable for rendering large drawings.

Markers tend to 'bleed' (where the ink spreads over the outline of drawing) when used on ordinary paper. Special 'bleed-proof' papers are available and should be used if you are going to use this type of rendering.

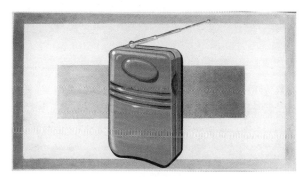

Figure 3.20 Marker pen rendering

The airbrush

Figure 3.21 An airbrush

Airbrushes are small spray guns that are used to spray liquid inks, dyes or paint onto a drawing or model. They are usually used for high-quality illustrations, such as final presentation drawings for a design concept. Airbrushing takes time and requires patience and practice. Areas of the drawing that are not to sprayed need to be 'masked', using masking film. This prevents colour being applied to wrong parts of the drawing through 'overspraying'. The airbrush can be used to provide flat tone over a large surface, thin lines or graduated tones. It is very important to practise your technique with the airbrush before starting the rendering of your final drawing.

Dimensioning

Your final drawings of your developed design should provide details of the size of each part of the product. The method you use to present the drawing is entirely your choice. Complex designs may require different views of the object to ensure that the dimensions are clear and easy to understand.

You should consider the following rules when adding dimensions to your drawing:

- The dimensions should be spaced well away from the drawing to prevent confusion

- All dimensions should give the 'real life' size of the object in mm.

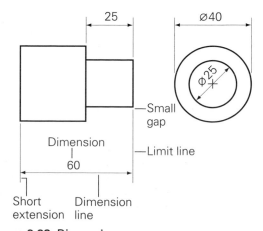

Figure 3.22 Dimensions

- Dimensions should be positioned so that they are the correct way up when read from the bottom or right-hand side of the sheet.
- Dimension lines with small arrow heads and limit lines should be used.

- The dimension should be written above the arrow head.
- It is important that the dimension lines do not get confused with your drawing lines.

3.4 COMMUNICATING DESIGN IDEAS – ANNOTATION

Fully communicating your design thinking and decisions to others is an important aspect of the work of the designer. In your own work, it is important that you use a range of techniques to clearly communicate your design idea. A number of methods exist to enhance design communication. The most important of these is 'annotation'.

To communicate effectively, you need to know the difference between labelling and annotation. Identifying features of your design through labelling items (such as 'pocket', 'flap' or 'screen') is the most basic form of annotation. These features will usually be clear to the person looking at the drawing and therefore labels will add little further information.

You should use annotation to communicate your design thinking clearly. While designing, you may think of different methods of assembly or operation of parts of your design. Through the use of brief notes, you can explore these alternatives and (this is important) communicate them to your teacher and the examiner. It is important to remember that the examiner will not meet you and will therefore not be able to question you about your design. You need to fully communicate your design skills and thinking through your design drawings and your annotation.

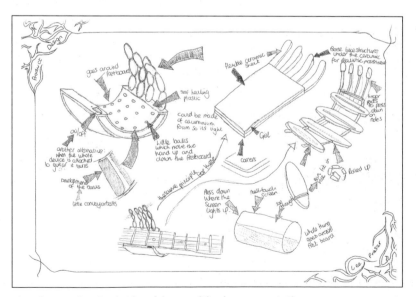

Figure 3.23 Example of a student's design ideas with clear annotation

CHAPTER 4

PRODUCT DESIGN TOOLBOX – MODELLING

By the end of this section, you should have developed a knowledge and understanding of:

- the importance of modelling in the design process
- the different types of model and how they are used by designers
- a range of techniques for making successful models.

4.1 WHAT IS A MODEL?

A model is something that is made to help the designer to test or evaluate an idea without having to make the product. A model can be a simple paper structure, a detailed prototype made from card and plastic and wood, such as balsa and jelutong, or an image produced using computer software.

The main function of a model is to help give the designer a better understanding of what the final product may look like. It is difficult to decide on whether a design is acceptable or not just by looking at a drawing. A full-scale model allows the designer to carry out tests and evaluations without having to go through the expense of making the real thing.

For example, motor manufacturers make a full scale model in clay to evaluate the appearance and function of a new car design.

Figure 4.1 Model of car in clay

4.2 CONCEPT MODELLING

It is here that ideas or concepts can be quickly tested for shape, form and functional details.

Drawings of your design ideas are useful for looking at shape, but the ideas can only be developed by understanding what the product will actually look and feel like. This can be achieved by making very basic concept models from styrofoam.

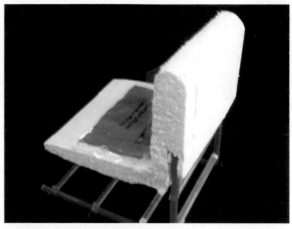

Figure 4.2 A foam model

4.3 TESTING THE FUNCTION

Sometimes when we are designing a product to work in a particular way, models can be used to simulate the mechanisms or movements that are needed. Test models don't necessarily have to look like the real thing, as long as the working part of the model is realistic.

Figure 4.3 Models help designers see how a mechanism may work

4.4 COMPUTER MODELLING

Computer models are a quick and inexpensive way of looking at the three-dimensional form of a product. The image of a product can be created on a computer monitor, and realistic colour and shading applied. Sophisticated computer programs allow these images to be rotated so that they can be seen from different angles.

Making changes

The main advantage of computer modelling is that changes can be made on screen. The

Figure 4.4 A 3D image is a quick way of seeing a product's form

images can be changed quickly, saving time and money.

Computer models allow the designer to make changes to the design and immediately see the impact it will have on the product.

Modelling for evaluation

When carrying out an evaluation on a product it is useful to compare your design with similar products. Computer modelling can help by making a **virtual** catalogue.

Figure 4.5 Virtual catalogue

4.5 MODELLING MATERIALS

The most common materials used for modelling in schools are medium-density fibreboard (MDF), cardboard and thin plastic. Fillers, paints, Plaster of Paris and other materials can also be used.

Medium-density fibreboard (MDF)

MDF is made from wood dust that is pressed together, glued and formed into large boards.

It is widely used for making small product models. Because it is a manufactured board and not a natural timber, it does not have any grain. This means that it does not split easily and can be formed into quite intricate shapes.

KEY POINT

- Because MDF is made from fine dust and glue, care must be taken and an extractor used when it is being machined

Expanded foam

Designers use expanded foams such as polystyrene as concept models. A concept model does not have any detail but it helps to give an impression of the shape, form and size of a product. Expanded foams are very crumbly and therefore do not allow any precise details to be formed.

Figure 4.6 Student using tools to shape MDF

Figure 4.7 Foam crumbles very easily but it is useful as a concept model

Foam board

Foam board is made from three layers of material. The outer surfaces are made from good quality white cardboard, and sandwich a thin layer of foam. Foam board is very light and can be glued together to make simple models of buildings and rooms.

Figure 4.8 Structure of foam board

Cardboard

Cardboard is probably the most widely used material for modelling. It can be made to look realistic by the addition of surface papers. For example, brick and tile patterns can be printed off from a computer and added to make a model of a house look very realistic.

Figure 4.9 Model of building with brick and tile papers

Plastic

Plastics such as acrylic and polystyrene can be used to make representational models. Acrylic can be heat formed or used as a flat sheet. It is particularly useful in modelling because it is so easily formed.

Figure 4.10 Modelling using plastic

Clay

Clay can be very useful to graphic designers. It is easy to use and bits can easily be added or taken away. It takes great skill to produce a very detailed model but clay can be useful for testing concepts such as ergonomics. For example, squeezing clay in your hand can give you an idea about the best ergonomic shape for a handle.

Clay is also very useful for making moulds for vacuum forming; it is simple and quick and allows textures to be added.

4.6 FINISHING MODELS

To be most effective, models must always look realistic. A wide range of different effects can be achieved with a little thought and imagination.

Textures

To make realistic textures a tin of spray paint and a few ideas are all that is needed. For example, sawdust sprinkled onto a glued surface can be made to look like grass when sprayed green. Spray glue added to the surface of a model feels like grip when it has been oversprayed with paint. Paper circles can be added to a model to give textured patterns. Sweets such as Smarties and Tic Tacs make very good buttons for model calculators.

CHAPTER 5

PRODUCT DESIGN TOOLBOX – USING ICT

5.1 INTRODUCING POWERPOINT®

By the end of this section, you should have developed a knowledge and understanding of:

- how to use PowerPoint® as a vehicle to present your work
- how to compress the size of image files
- how to package your PowerPoint® file, enabling it to be saved to CD
- how to insert sound and movie files
- how to use hyperlinks to link different elements of your work together.

Before the development of computers, design proposals were always presented using a series of display boards. In some cases this still happens today. However, with the increasing development of IT, computer-based presentations are increasingly popular.

Whichever methods you use, it's important to consider the design of your slides and the audience who will need to view them. They need to be clear, yet stimulating.

PowerPoint® is similar to word-processing software like Microsoft® Word, except that it is geared towards creating presentations rather than documents. Word documents consist of one or more pages, and PowerPoint® presentations consist of one or more slides. It is easy to delete slides, rearrange them if you need to and modify or add to them. So when it comes to keeping a production log, for instance, it is easy to continually add photographs and update your work as you progress. You can use the program to do this for either electronic or paper portfolios; you will just need to print off the information for inclusion in the paper portfolio rather than present it electronically.

▶ PowerPoint® – keeping it simple

Using PowerPoint® as a tool to present your work has its advantages:

- It's good for the environment, as you do not have to print out work. This saves paper, ink and electricity.

- You can add sound, video and animation, which you can't do if you produce a paper portfolio.

However, there are disadvantages:

- If you are not careful you could lose your work.

- PowerPoint® has lots of clever techniques for making slides appear, such as text that rotates. It's easy to get distracted by them and then not provide the evidence the examiner needs in your portfolio. Remember to keep it simple.

EXAMINER'S TIPS

Use a plain white background with your work inserted for your portfolio. Avoid the use of special effects. Remember the examiner is looking at the content you provide within your portfolio and not the visual effects you use.

Remember that you are not creating a PowerPoint® presentation. You are creating a controlled assessment portfolio on a computer. The examiner will therefore expect lots of information on each slide. This may include: sketches, notes, tables, etc. The basic rule is that you treat each slide in the same way you would present on paper: the folio is just presented electronically.

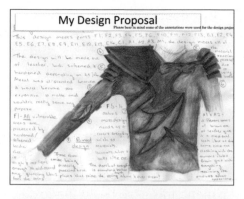

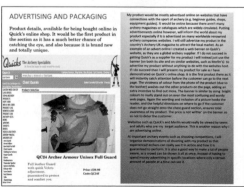

Figure 5.1 Good examples of Powerpoint® slides; each contains lots of information

All your work does not have to be on PowerPoint®. It could be on Word, a drawing package, scanned images or photographs. These can be easily linked to your presentation using hyperlinks (we will talk about hyperlinks in more detail later).

If you wish, you could do all your work using Word and use PowerPoint® to link the pages together – like putting pages into a folder, PowerPoint® in this case being the folder.

KEY POINT

- View PowerPoint® as a folder that you put different pages into. Think 'clear and simple'. There is no need to waste time using fancy PowerPoint® gimmicks. It is a tool to present your work electronically.

▶ Essential tips for using PowerPoint®

This section will help you to get the best out of PowerPoint®.

Compressing the size of image files

As your electronic portfolio is going to contain a lot of photographs, it is a good idea to compress the size of these images so your file doesn't get too big to manage. It is easy to do this.

Start by double clicking on any image in your presentation ⇨ Select Compress picture icon.

Figure 5.2 Compress picture icon – screenshots

A dialogue box will appear. Select Options, and another box will appear. Check Screen (150ppi): good for web pages and projectors. Select OK in both boxes.

Figure 5.3 Screenshot – Compress picture dialogue box

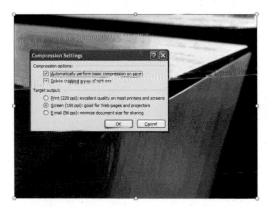

Figure 5.4 Screenshot – Compress settings

PowerPoint® will now compress all of your images. This will significantly reduce the size of your presentation, making it easier to copy and manage.

Packaging your presentation on a CD

This is essential, to make sure that the examiner receives all the information you have entered into your electronic portfolio.

One of the features of PowerPoint® is the

Package for CD option. If your computer has a writable CD drive, you can use this option to create a self-contained CD, which contains your presentation along with any necessary supporting files (such as fonts, sound files, video clips, images, etc). It will also create PowerPoint® Viewer, which lets you view the presentation on any computer, even if the computer does not have PowerPoint® installed.

To create a CD with your presentation, follow these seven steps:

1. PowerPoint® 2003, choose File ⇨ Package for CD or

   PowerPoint® 2007, select the Office button ⇨ Publish ⇨ Package for CD.

Figure 5.5 Screenshot – Office button drop-down menu

Figure 5.6 Screenshot – Publish folder options

The Package for CD pop-up box appears.

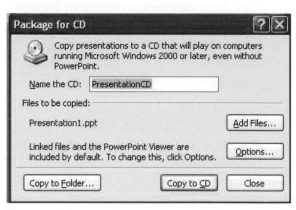

Figure 5.7 Screenshot – Package for CD

2. Click Options

   This opens the Options dialog box.

Figure 5.8 Screenshot – Package for CD options

3. Study the options and change any you require. The following points describe each of the options that are available:

   • PowerPoint® Viewer: ☑ Check this option to include the PowerPoint® Viewer on the CD. If you include the viewer, you can use the drop-down list beneath this option to indicate whether you want to play the presentations on the CD automatically or allow the user to choose which presentation to play.

- Use the drop-down menu and select 'Let the user select which presentation to view.' This is a much better option if your work is selected for moderation, as the person looking at your work will prefer to open your work as a PowerPoint® file.

- Linked files: ☑ Check this option to include any linked files, such as videos or large audio files.

- Embedded TrueType fonts: ☑ Check this option to ensure that the fonts you used in your presentation will be available when you show the presentation on another computer.

4. Click OK to return to the Package to CD dialog box.

5. Click Copy to CD. (Figure 5.7)

   If you haven't already inserted a blank CD into the drive, you are prompted to insert one now.

6. Insert a blank CD into your CD-RW drive and click Retry.

   PowerPoint® copies the files to the CD. This may take a few minutes. After the CD is finished, the drive will open and ask if you want to make another copy.

7. Remove the CD, and then click Yes, if you want to make another copy. Otherwise, click No, and then click Close.

The process is now finished.

Inserting a sound object

This section explains how to insert a sound object onto a slide. You can configure the sound object to play automatically whenever you display the slide, or you can set it up so that it will play only when you click the sound object's icon. If you want to control when the sound plays, or if the sound file is in a format other than WAV, follow these six steps:

1. Move to the slide to which you want to add the sound.

2. PowerPoint® 2003, choose Insert ⇨ Movies and Sounds ⇨ Sound from File.

   PowerPoint® 2007, choose Insert tab ⇨ Select the speaker icon

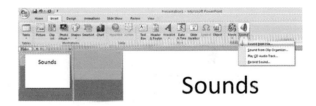

Figure 5.9 Screenshot – Insert sound icon and drop-down dialogue box

3. The Insert Sound dialog box appears.

4. Select the sound file that you want to insert.

   You will have to browse your computer files to find the folder that contains the sound files you require.

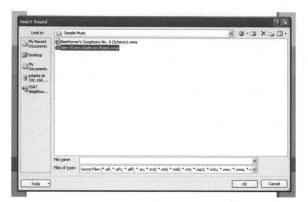

Figure 5.10 Screenshot – Insert sound from file

5. Click OK.

A dialog box appears, asking if you want the sound to play automatically when you display the sound, or play only when you click the sound's icon.

6. Select Automatically or When Clicked.

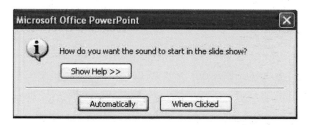

Figure 5.11 Screenshot – How do you want the sound to start in show? – Dialogue box

The sound clip is added to your slide and represented by a little speaker icon.

Figure 5.12 Sound icon will show on slide when inserted

However, if the sound is a WAV file and you want it to play automatically, it is easier to add it to the slide transition than to add it as a separate object.

EXAMINER'S TIP

The process for inserting a video clip is the same as the process for a sound clip.

Creating a hyperlink

Hyperlinks aren't limited to PowerPoint® presentations. In PowerPoint®, you can create a hyperlink that leads to other types of Microsoft® Office documents, such as Word documents or Excel® spreadsheets. When the person viewing the slide show clicks one of these hyperlinks, PowerPoint® automatically runs Word or Excel® to open the document or spreadsheet.

Adding a hyperlink to a presentation is easy. Just follow these steps:

1. Highlight the text or graphic object that you want to make into a hyperlink.

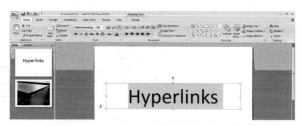

Figure 5.13 Screenshot – Highlight the item you want to act as a hyperlink

The most common type of hyperlink is based on a word, or words in a slide's text area, alternatively it could be an image.

2. Choose Insert ⇨ Hyperlink.

Alternatively, click the Insert Hyperlink button found on the standard toolbar or use the keyboard shortcut Ctrl+K. The Insert hyperlink dialogue box will appear.

Figure 5.14 Screenshot – Select hyperlink icon

3. Click the Place in This Document icon in the list of four icons on the left side of the Insert hyperlink dialog box.

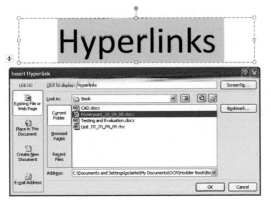

Figure 5.15 Screenshot – Insert hyperlink dialogue box

The four icons are as follows:

- Existing File or Web Page: This means you can link to another file in another application, or to a web page on the internet.

- Place in This Document: This means you can link one part of your PowerPoint® presentation to another part.

- Create New Document: This is just what it says it is. You can, however, choose now or another time to edit the new document by clicking the appropriate button.

- Email Address: Use this to link to an email address. This feature is useful in an intranet or internet setting because the link allows the reader to write email to the email address that you link to.

If you click the Existing File or Web Page icon, you can then select your link from the following:

- Current Folder: Allows you to choose any page in the current folder.

- Browsed Pages: Allows you to choose any page that you have browsed using your web browser recently.

- Recent Files: Enables you to view recently used files.

After you've found your desired item, and clicked it, click OK. And there you have your hyperlink.

If you want to link one part of the presentation to another part in the same presentation:

- Click the Place in This Document icon and the text box to the right displays the list of slides in your document. As you click each slide, you can see a slide preview.

- Click the slide that you want the hyperlink to lead to, and then click OK.

You return to Normal View. The Insert Hyperlink dialog box vanishes, and the hyperlink is created.

If you create the hyperlink on text, the text changes colour and is underlined (usually in blue). Graphic objects such as AutoShapes, WordArt, or clip art pictures are not highlighted in any way to indicate that they are hyperlinks. However, the mouse pointer always changes to a hand pointer whenever it passes over a hyperlink, so providing a visual clue that the user has found a hyperlink.

Figure 5.16 Screenshot – hyperlink shows blue and underlined

5.2 COMPUTER-AIDED DESIGN (CAD)

LEARNING OUTCOMES

By the end of this section you should have developed a knowledge and understanding of:

- the role of CAD within the design process
- the difference between 2D and 3D CAD packages
- the advantages of using CAD within your design portfolios
- the use of CAD for modelling and testing.

The development of your skills in the use of computer-aided design (CAD) within the product design course is very important. Computer-aided design is widely used by designers in industry as a tool in the design and development of products. The work you will undertake as part of the course will allow you to develop your skills and use of both 2D and 3D CAD software packages.

The advantages of CAD

The use of CAD in the design and development of your prototype product offers you, as the designer, many advantages, including:

- the ability to make alterations to components quickly, without the need to redraw the whole component
- the opportunity to examine your design through the use of rotation, assembly, zoom
- the opportunity to render the product to show different colours/materials

- the opportunity to test your design using features such as 'finite element analysis', which allows you to predict the effects of 'loading' and find weaknesses in your design, allowing alterations to be made before production

- the ability to output your design to computer-aided manufacturing (CAM) systems that will allow you to both model the manufacturing process on screen and ultimately to manufacture your design

- the ability to produce working prototypes using rapid prototyping systems.

Types of CAD systems

The type of CAD system or systems that schools and colleges use varies from one school or college to another but the usual features include:

- desktop or laptop computer
- 2D and/or 3D CAD software
- graphics tablet
- scanner
- digital camera
- internet access.

2D CAD software packages

Using a 2D CAD package will allow you to produce drawings without the need for extensive training. The tools are intuitive and you can develop CAD drawing skills quickly.

Software such as Techsoft's 2D Design V2 will allow you to produce many different styles of drawing to aid both your design communication and design modelling, and, if appropriate, undertake the subsequent manufacture of your product.

Figure 5.17 Techsoft 2D Design V2 Drawing – Dragon Menu

The software makes it possible to do three things:

- to undertake graphic design activity – allowing you to combine vector graphics, text, bitmap images, photographs and clipart to create and manipulate images for items such as logos, menus, point-of-sale display and product packaging

- to produce detailed 'technical drawings' – the software allows you to use a range of tools to produce accurate, dimensioned drawings

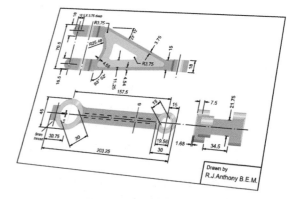

Figure 5.18 Dimensioned working drawing

- create manufacturing drawings that can then be outputted to a variety of CAM devices such as vinyl cutters,

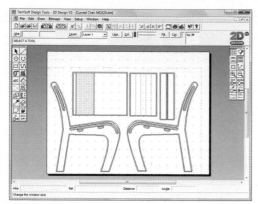

Figure 5.19 Manufacturing drawings for chair

plotter/engravers, millers, routers and laser cutters.

3D CAD Software Packages

Using 3D CAD software allows designers to generate three-dimensional images of their design ideas quickly.

EXAMINER'S TIP

The software is not just a method of producing a 'lifelike' visual image of the product – it also aids product modelling and design development. It is this use of the software that examiners will be looking for if you are to score highly in the controlled assessment task for Unit A551, Developing and Applying Design Skills.

The type of 3D CAD software that is used varies from one school or college to another. Software such as 'Solidworks' will allow you to produce full 3D models of your design quickly. The software will allow you to analyse designs and compare different versions to obtain the optimum design for your product.

You can test designs using 'finite element analysis', which will 'load' your design, allowing you to identify areas of weakness prior to manufacture.

Other software packages, such as SpeedStep for textiles, offer alternative approaches to the use of CAD. The real value of Speedstep for students is the ability to design fabric and put it into repeat, which then can be digitally printed. You can use drawings that you have done yourself – either scanned in from sketchbooks, drawn in ProSketch from the object library or using the templates provided in ProPainter to create the fabric design. You can consider fabric colours, patterns and textures by 'draping' them on photographic

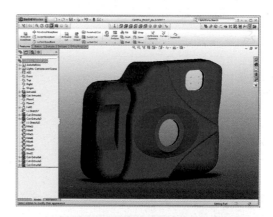

Figure 5.20 Solidworks camera design

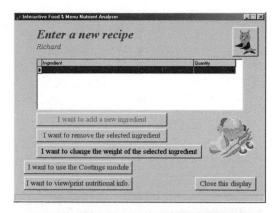

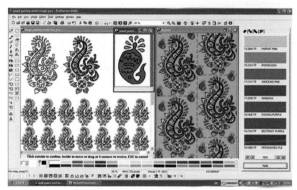

Figure 5.21 Screenshots of SpeedStep and Prosketch

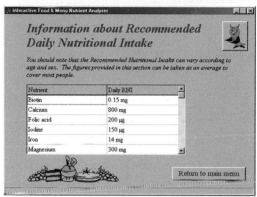

Figure 5.22 Screenshots of 'Birchfield' nutritional and sensory analysis software

images, allowing you to assess the design before printing it onto fabric.

For food products, nutritional analysis and sensory analysis software is available.

5.3 COMPUTER-AIDED MANFACTURE (CAM)

LEARNING OUTCOMES

By the end of this section you should have developed a knowledge and understanding of:

- what CAM is and what it stands for
- the CAM machines that are available and what they can do
- the advantages and disadvantages of CAM.

Today, an increasing number of production processes can be carried out by machines that are controlled by computers. Computer-aided manufacture (CAM) is the term used to describe the process. Automated manufacturing is quicker, it can run 24 hours a day, and it is safer, more reliable and cost effective in the long run.

The advantages of CAM

One of the main advantages of CAM is that files can be stored electronically and downloaded when needed. These instructions can be sent to the equipment via the computer very quickly and small changes can be made to designs very quickly, when needed. Often, products are not made entirely in one place. Components are manufactured in separate locations and then brought together to be assembled. The electronic storage method also means that information can be shared quickly with people all around the world – wherever manufacturing is taking place.

Where CAM has replaced a manual operation, it gives greater accuracy and precision. Greater productivity can be achieved, as the CAM machine does not need to take a break. CAM is also safer when using materials and chemicals that could be harmful to us.

Most products manufactured today are made using a combination of manual and CAM operations.

Computer-integrated manufacture (CIM)

CIM is a totally automated production system, where the whole production process is computer controlled. The development of powerful CAD/CAM systems has allowed the entire design, production and manufacturing processes to be controlled by a single computer. Companies who have these facilities are able to make dramatic price reductions, while increasing the quality and reliability of their products.

Types of CAM equipment

Pieces of CAM equipment in simple terms are things that are linked to computers that output information from the computer to produce things. The most common of these is a printer, which enables the production of graphic work, including photographs.

Knife-cutting machine

This type of machine uses a sharp blade to cut through thin material, such as sheet materials, including card and vinyl. The ability to cut these materials accurately allows products to be made in quantity and to a high standard of finish.

Laser cutter

The laser uses a beam of intense light to burn through the material. It has to be focused carefully on to the material. Laser cutters can cut and engrave sheet materials. They can cut a range of plastics very accurately, leaving a smooth finish. They also cut wood, MDF, paper, card, cork, cloth, laminates, circuit boards and even chocolate! They can engrave or mark all these materials and others, such as glass, slate, fabrics and ceramics. Fumes are given off when cutting some of these

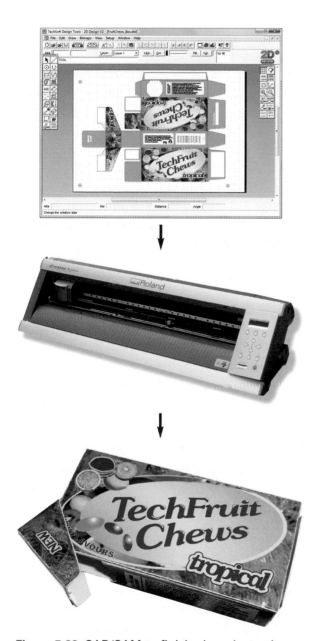

Figure 5.23 CAD/CAM to finished product using card to create packaging

Figure 5.24 CAD/CAM to finished product using acrylic to create a picture frame

materials, so the laser cutter must have appropriate extraction facilities.

For students and teachers, the real advantage of the laser cutter is the minimal set-up and clean-up required, and the speed at which large jobs can be processed.

Milling machines and routers

These machines use stepper motors to move a cutting tool in three or four axes. A three-axes machine has X, Y and Z axes, enabling 3D shapes to be manufactured. A four-axes machine allows the work to be rotated, so that the shape can be cut in one operation. Computer Numerical Control (CNC) milling and routing are an essential strand of modern manufacturing practice.

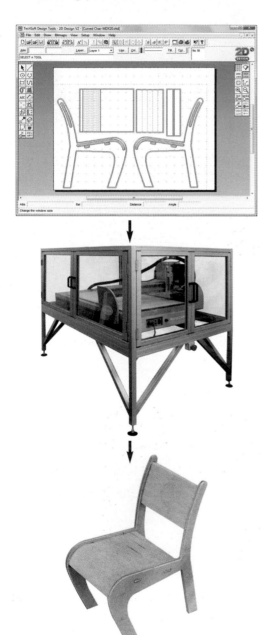

Figure 5.25 CAD/CAM to finished product using plywood to create a child's chair

Figure 5.26 Guitar body cut and shaped by CNC milling machine

Rapid prototyping

Rapid prototyping is 3D printing, where a product is built up and produced in a series of layers. Sophisticated CAD programmes can produce stereo lithography files (STL files), which slice a design into very thin layers.

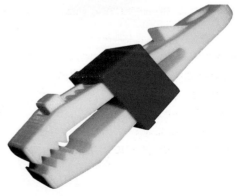

Figure 5.27 Techsoft 3D printer and products

Figure 5.27 continued

Figure 5.28 Computerised embroidery

These layers are then built up to create a 3D shape.

If you have the correct file, you can produce a model of yourself or your skeleton, but it would take some time to achieve as these types of processes can be slow, depending on the intricacy of the design. However, they are accurate.

The more common rapid prototyping machines use lasers to cure a liquid resin a layer at a time. Other systems extrude a tiny thread of molten plastic to build up the layers. Both of these processes are time consuming and can take several hours to create a simple part. However, at the end of the process a solid plastic model is built up, which can be handled and even used as a real working part.

Computerised embroidery

These machines are used widely in the textiles industry and allow designs to be represented by a series of stitches in a series of colours to build up a design. They are used to produce a quality product that can be replicated on a large scale if required.

Summary

The advantages of CAM:

- continuous and quicker production
- less labour intensive, so production is cheaper
- improved consistency in quality
- machines can be programmed quickly to perform different operations.
- ability to send files anywhere in the world to be manufactured.

The disadvantages of CAM:

- data can be lost if not properly stored and backed up
- reduces the need for a labour force, cutting job opportunities
- computer technology is expensive to buy and maintain
- usually requires a large investment to train staff to operate the CAM systems.

CHAPTER 6

PRODUCT DESIGN TOOLBOX – TESTING AND EVALUATION

By the end of this section, you should have developed a knowledge and understanding of:

- the importance of relevant testing in context
- how to gain relevant information from your user group
- how to evaluate the success of your specification
- how to use your findings to improve and modify your product.

6.1 TESTING AND EVALUATING YOUR DESIGN

Testing and evaluating are an integral process in the world of design and are undertaken all the time. For example, you might be required to:

- analyse and evaluate your research findings to help you make progress,

reflecting upon your results to help you find new ways to work and solve problems

- evaluate an existing product to identify strengths and weaknesses in its design and manufacture, with a view to making improvements
- evaluate your designs and models to help you make the right decision about a product's development – this could include the materials, fabrics or ingredients that you choose, the way in which you might

As you work through the course you will often carry out testing and evaluation. As a designer you will need to find out how successful your designs and products are by testing them with potential users. How well does your design match your design specification? What might you do differently if you could start again? How might your design be improved or modified?

KEY POINT

- It is important that any testing or evaluating you undertake refers directly to your specification and to your target user group.

combine them and the way in which you might finish them off

- evaluate the best way to manufacture your models and products – to ensure that resources, materials, fabrics and time are used economically and in the best way possible

- test and evaluate the things you make – to help you make your future products even better.

6.2 TESTING YOUR PRODUCT DESIGN SOLUTION

You must test a design to ensure it is successful. Ways to do this include:

- asking your user group what they think about it. You could formulate a simple questionnaire and give it to your specific users. Your questions should directly reflect your specification.

- finding out if it does what it is supposed to do. Use your product for the purpose for which it has been designed and record how well it functions.

- giving your own critical thoughts on your solution. Honesty is the best policy. Be honest about your product but suggest ways in which you could make it better. Imagine that you are the customer giving feedback on the design.

- testing your solution in context with the identified target users of the product. Take the product to the environment it is designed for and test it with your target group. Does it do what you need it to do? You will need to plan this carefully – you will need time to organise people to test your product.

- asking experts for their opinion. Getting feedback from people who really know what they are talking about is a great idea. You may find that local businesses are happy to help. You could ask your teacher to support you in making contact with them directly. You may be able to arrange to demonstrate your product to them and ask their opinion.

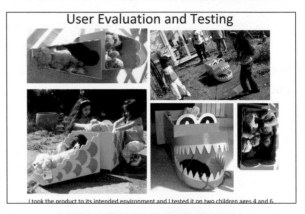

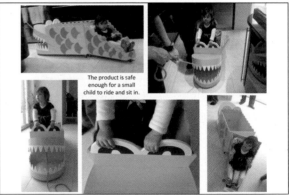

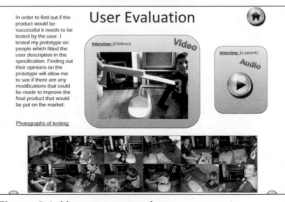

Figure 6.1 User group testing

EXAMINER'S TIP

These images show the product has been tested in context. The work would be improved if the images were annotated with thoughts and findings. If you are submitting your work electronically, you can insert video and sound clips into your portfolio. This could be very useful for user testing.

ACTIVITY 1

Look at the five everyday products shown here. Identify a way that you could test each one.

Figure 6.2 Everyday products

6.3 EVALUATING YOUR PRODUCT DESIGN SOLUTION

The success of your design usually depends on whether it meets your specification and whether your user group likes it. In your evaluation, comment on:

- how well the design/product meets the needs of your user group. What did they need your design or product to do? Does it

fulfil their needs and requirements? Do they have any difficulties in using it?

- how well it works. Is your product reliable? Will it work the same way every time it is used?

- whether it functions as you intended it to. Does it do what you designed it to do? Are there any surprises that have inspired you to develop it or enabled you to seek other possibilities for your designs?

- what your product looks, feels, smells, tastes and sounds like. Have you achieved a pleasing design that appeals to the senses?

- whether the materials you used were appropriate. Were you able to use sustainable or recycled materials in your products, could you have found alternative materials that might have been cheaper, kinder to the environment or more appropriate to the task?

- whether the manufacturing processes were appropriate. Have you been able to demonstrate the use of the most sensible methods of manufacture? Did the methods you chose enable you to demonstrate your skills to their full potential?

- whether it could easily be manufactured in quantity. What opportunities were there for you to demonstrate the use of batch production, repetitive flow and continuous production? How exactly would your product be made in quantity in an industrial context? What factors might you need to consider when making your product in quantity?

- whether ICT and CAD/CAM could be used effectively to manufacture the design. What opportunities are there for you to use new technology to support the manufacture of your design?

Your evaluation needs to be critical. Nothing will get better if you don't challenge the design and look for alternative solutions.

The following images are examples of students' work showing different ways of evaluating, testing and modifying their work.

The evaluations of the specifications are detailed and give justified feedback on each specification point.

The testing of the products is done in context and refers to their user group's experience of using the product. Evidence is shown using both images and detailed annotation.

The examples of how products are modified use images and clear explanations on how the products could be improved by making small changes to the design.

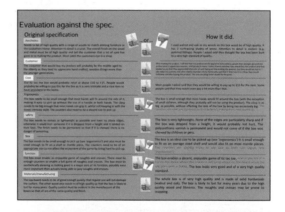

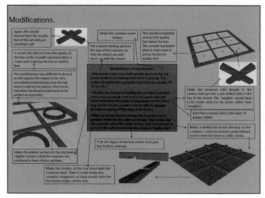

Figure 6.3 Student examples of testing, evaluation and modifications

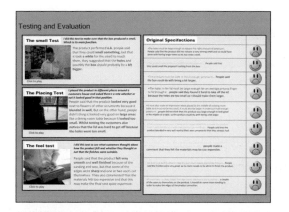

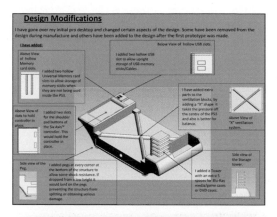

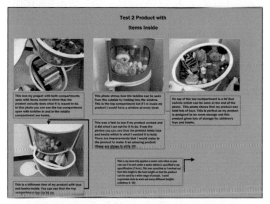

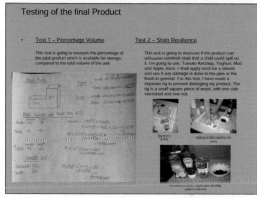

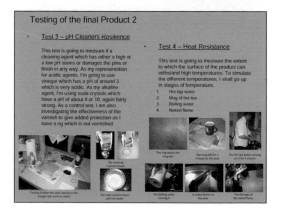

Figure 6.3 continued

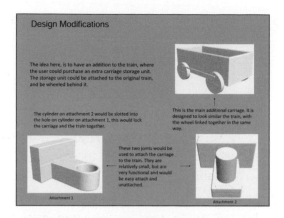

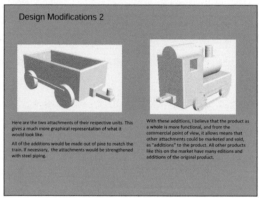

Figure 6.3 continued

KEY POINT

- You must justify all your responses. Explain your answers and show where the evidence came from. Don't be afraid to show both good and bad points about your work. Showing you can identify weaknesses in your solutions allows designs to be modified and improved, moving things forward.

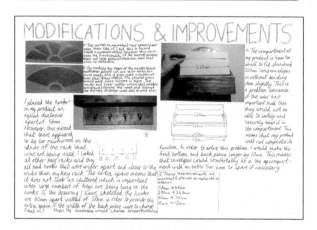

▷ It would be even better if …

Consider what you could do to develop and improve your work. Look at this as an opportunity to show your creative and innovative side.

When modifying and improving a product you should be able to let your imagination run away with you. No matter how unrealistic you think your ideas might be, it will give you the opportunity to 'think outside the box' and develop something completely new and different. Do not be scared of change – embrace it!

Figure 6.4 Good examples of modifications, showing sketching and annotation.

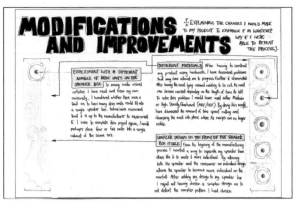

Figure 6.4 continued

Other questions to consider when modifying your designs and products

1. If the product is made overseas, how would you improve the working conditions of the workers?

2. Is the design likely to offend? How could you make it more acceptable for everyone?

3. How could you consider different cultures in the design and promotion of the product?

4. Could you make the design more socially acceptable and in tune with the current ideals and thoughts of our time?

5. Does the product take into account the needs of people who have a disability? How could you modify the product to enable a person who has a disability to use it?

6. What options are there for powering the design? Could you include natural elements of wind, water and solar as sources of power in the production of the product? How might you improve the way it is made?

7. Could you improve the product to enable the user to interact with it more comfortably?

8. What opportunities are there to develop the look, taste, touch, sound and smell of the product?

9. What laws might you need to develop in order to protect the user and the designer of the product?

10. What parts of the product could be reused, recycled, reduced and repaired?

11. What could you do to help promote the idea of sustainability?

12. How could you develop the product to help support different scales of production?

13. What opportunities are there for the use of CAD/CAM in the product?

14. How could you improve the product to ensure the health and safety of the user and the product?

15. Could you change the materials, fabrics or ingredients of the products to make them better?

ACTIVITY 2

A further five everyday products are shown. Sketch how you might modify and improve each one. You should consider the relevant 'design influences' when thinking about improving or modifying these products. As you look at each product, think carefully about the questions listed above.

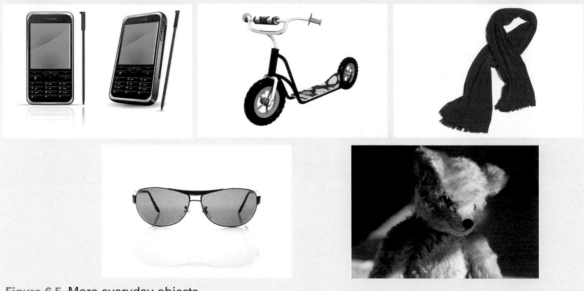

Figure 6.5 More everyday objects

CHAPTER 7

PRODUCT DESIGN TOOLBOX – DESIGN INFLUENCES

By the end of this section, you should have developed a knowledge and understanding of:

- the 'design influences' and how these influence the design of products
- the effects of the design influences on real 'case studies'
- how the design influences can be interrelated
- how to apply your knowledge and understanding of the design influences to your own designing
- how to relate your knowledge and understanding of the design influences to existing products

The 'design influences' that are detailed below are central to product design. Designers will consider each of the design influences during the design of products. Some will have more influence than others depending on what is being designed, who the product is for and where the product will be used and marketed.

7.1 THE DESIGN INFLUENCES

Each design influence should be considered equally when designing products. Occasionally, it is possible that a specific design influence will have no relationship with that product at all. A piece of one-off bespoke jewellery for example would not need to consider the influence of Computer Aided Design and Manufacture.

The design influences used for GCSE Product Design are:

- Social, moral and cultural issues
- Environmental factors
- Ergonomics & Anthropometrics
- Aesthetics
- Patent and copyright and consumer law
- Computer Aided Design and Manufa

- Marketing and economics
- Colour theory
- Systems and structures
- Energy
- Scientific principles
- Health and safety
- Globalization of design and manufacturing
- Sustainable technologies
- Sustainable design imperatives and design tools
- Economics of manufacturing

Social, moral and cultural issues

When we design products we should always look beyond our own lives and experiences. We live in a diverse and culturally rich country. It is not only important to learn from different cultures, but also to understand them in order to avoid offence and mistakes. It is important for us as designers to recognise the potential that products have to offend. As designers, we have a social and moral duty to respect other people's beliefs and cultures.

Even when we don't offend, it is easy to make a mistake if our research into different cultures, traditions and languages is not thorough enough. One of the most common mistakes made by Western companies is the failure to translate accurately the language of the country to which they are importing their products.

Apart from avoiding such blunders, as designers we can learn a lot from researching different cultures. For example, Islamic art is very different from Western art. Within the majority of Islamic traditions, using images of

CASE STUDY

One famous drinks company from the UK once translated its slogan, 'Turn it loose', into Spanish, where its translation was read as 'Suffer from diarrhea'. Naturally, this product wasn't as popular as they'd hoped.

In the same way, a company famous for manufacturing hair straighteners decided to market their product the 'Mist Stick' in Germany. Unfortunately, in German 'mist' is slang for manure. Not too many people had use for the 'manure stick'.

When a large food production company started selling baby food in Africa, they used the same packaging as they did in America, with a beautiful baby on the label. Later they learned that in certain African countries, companies routinely put pictures on the label of what's inside, since most people can't read.

An American T-shirt maker in Miami printed shirts for the Spanish market, which promoted the Pope's visit. Instead of 'I saw the Pope' (el papa), the shirts read 'I saw the potato' (la papa). Oops!

people is not permitted, and instead patterns made of geometric designs are used. Complex geometric designs create the impression of continuous repetition, which is believed by some to represent the infinite (unending in time and space) nature of God. In mosques and in other buildings, as well as on common objects like bowls and rugs, geometric designs are very common. These patterns can form the inspiration for a wide range of products.

In Islamic design, the star is a regular geometric shape that symbolizes equal radiation in all directions from a central point. All stars are created by dividing a circle into equal parts.

Figure 7.1 The five-pointed star repeats in patterns based on a square

▶ Environmental factors

Environmental factors have obvious and direct connections with aspects of some of the other design influences such as 'sustainable technologies' and 'sustainable design imperatives'. The design influence 'environmental factors' is not just about pollution, recycling and using renewable sources of energy.

Depending on the products being designed, consideration of such things as the weather, wind, heat, humidity and cold can also have an effect on how a product looks and how it performs.

The section of this chapter called **efficient environment**, talks about the ambient working temperatures that enable us to function at our best. The temperature can also effect the correct function of products

CASE STUDY

One very famous example of how cultural traditions led to a world-famous design is the design of the 'Sony Walkman', the forerunner of all portable CD/MP3 players. Within Japanese culture, it is hugely important for people to have their own personal space; like a clear bubble around them that no one else can enter. However, Japan is one of the most crowded countries in the world, and likewise, Tokyo is an overcrowded city. It is said that Nobutoshi Kihara, the designer of the original Sony Walkman, gained the inspiration for the groundbreaking new product while feeling extremely uncomfortable being hemmed in on a Tokyo underground train. Desperate to regain his personal space, he dreamt up the idea of the portable tape player – a simple idea that fills your head with music, and therefore cushions against your immediate environment. This concept has also been widely adopted in the West as environments have become more crowded.

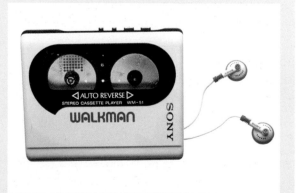

Figure 7.2 The Sony Walkman developed from a cultural need

too. Low temperatures can cause pipes to freeze whilst high temperatures can result in equipment such as computers failing. Consideration of materials, components and temperature control as part of the design would overcome these types of problems.

Light is essential to our lives and our enjoyment of many products. The use of photo-electric cells, infra-red devices and light dependant resistors (LDRs) in products makes them more efficient, safer and comfortable for us to use. However, light does not always have a positive effect on products. It can cause deterioration in the properties of certain plastics. This could result in product failure if not fully taken into account by designers.

Sound can please us, for example when we listen to music, but can also cause discomfort, such as the sonic 'boom' when an aircraft passes through the sound barrier. Designers need to consider the impact of sound when designing products. Sound can

also be used as a manufacturing process. It can be used for welding metals and plastics together.

Ultrasonic welding of plastics and metals and even some fabrics is a relatively modern assembly method. Bonding materials together in this way does not require adhesives or other consumables such as solder or thread. The 'weld' or bond is produced quickly – usually in two to four seconds – and without heating the surrounding area.

As designers we need to consider not only the environment around us but also the type of environment that the product will be used in. We need to consider our impact on materials and resources and ensure that we use materials in a sustainable way. By considering the environment in which the product will be used we can ensure that the product will operate correctly and will have a long service life, reducing environmental impact by minimising product disposal and replacement.

Ergonomics & Anthropometrics

Ergonomics

Whenever you design anything that is going to be used by humans, you must consider ergonomics. Ergonomics is the study of how people work (or rest or play) in their environment, which could be an office, school, factory or at home. In simple terms, ergonomics is about how to help people be more efficient at what they do.

A good example of how a product is ergonomically designed is a mobile phone. The phone has rounded edges to make it comfortable to use, the distance between the

Figure 7.3 Concorde, a regular breaker of the sound barrier

microphone and the speaker fits the distance between the average adults ear and mouth, and the buttons are well spaced and easy to use. Notice also that the buttons use a bold typeface that is easy to read.

Figure 7.4 The mobile phone is well designed to fit the needs of the human user

The efficient environment

Ergonomics is about making things the right shape, size and weight for humans. But what if the room that you are working in is too hot or too cold? People work best at 'room temperature' which is about 20°C. You cannot work efficiently if you are too hot or too cold. Ergonomics also considers noise, vibration,

Figure 7.5 Offices are carefully designed so that people can work efficiently

light, and smell. If your senses are uncomfortable you will not work efficiently.

Take a look at the picture in Figure 7.5. The person is sitting at a desk that has been ergonomically designed. The heating, lighting and noise are carefully controlled in offices so that people are comfortable and work at their best.

Ergonomic graphics

The sign in Figure 7.6 is written in an old English style.

Figure 7.6 The gothic script is very difficult to read

This style of very decorative writing is poorly designed from an ergonomic point of view. It is difficult to read and takes time to work out what it is saying. When designing signs and posters it is important that people can quickly and easily 'get the message'.

Text size

The size of text is very important. If the writing is too small, we have to strain our eyes to read it.

ACTIVITY

Look at the writing below. Which ones do you find difficult to read?

Can you read this?

Can you read this?

Can you read this?

Can you read this?

Can you read this?

Can you read this?

Anthropometrics

Anthropometric data is an essential resource for designers. The prefix 'anthro' means humans, and 'metrics' means measurement. Therefore, anthropometrics is about human measurement.

Anthropometric data sheets contain information about the range of human dimensions and sizes for all ages.

If you measured the hand width of 10,000 18-year-old men, you would find that the graph produced would follow a bell-shaped curve.

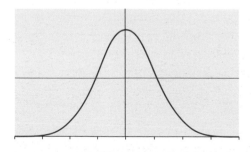

Figure 7.7 A normal distribution (bell curve) is formed when a large number of measurements are taken

At the left-hand side of the graph are people with small hands. At the right-hand side are people with large hands. The percentages help us to work out how many of the 10,000 people have small, large or average-sized hands. The vast majority of people are close to the average. Usually when we design things we try to consider 90 per cent of the population. This is the 90 per cent that lie between the 5th percentile (excluding the the bottom 5 per cent – those with small hands) and the 95th percentile (excluding the top 5 per cent – those with very big hands). Anthropometric data is produced in a series of tables that can be found in most libraries.

Style and colour

Some typefaces are easier to read than others. It is important that the one you choose can be read easily. This is particularly important for emergency signs, where clarity is very important.

Which style would you choose?

DO NOT ENTER

DO NOT ENTER

The way colour is used can greatly affect how efficient a graphic product is. Good ergonomic design requires colour contrasts, such as blue and white.

Nice and clear

Not so clear

Figure 7.8 Colour combinations are important. Always try to achieve contrast

▶ Aesthetics

'Between two products equal in price, function and quality, the better looking one will outsell the other.'

Raymond Loewy, industrial designer (1893–1986)

The word aesthetics is derived from Greek. It is a branch of philosophy that deals with the perception of beauty and taste. For designers, aesthetics means how we perceive (or see) beauty. As humans, we have five senses: sight, sound, taste, touch and smell. We use all of these senses when we have an aesthetic experience. Many students mistakenly think that aesthetics is only about how something looks. We often use the phrase 'aesthetically pleasing' without thinking about all the elements of a design that contribute to its beauty. For example, when you eat a beautiful meal, you see it, smell it and taste it – it is the combination of all three of these senses that give you the true aesthetic experience. However, it might be the packaging of something that first attracts you, so clearly appearance is hugely important.

Some people can be moved to tears listening to music, and others say, 'I like the feel of that material', but our perception of beauty is very personal. What one person loves, another person may hate.

The elements and principles of design

When considering aesthetics, designers need to draw on the elements and principles of design. These are the building blocks that are used to create products. Skilled designers draw on them in order to produce harmony. Harmony in product design means that all the elements work together to produce a visually satisfying product.

The main principles you need to know about for this qualification are outlined here:

- **Line** – Line can be considered in two ways. The marks made with a pen when drawing or, in product design, the edge created when two shapes meet.

QUESTION

1. Take a look at the two houses in Figure 7.9. Which would you rather live in?

To some people, the cottage is their idea of perfection, to others it is a nightmare.

Figure 7.9 Dream homes?

- **Shape** – A shape is a self-contained, defined area of geometric or organic form. A positive shape in a painting automatically creates a negative shape. Shapes are two dimensional, e.g. a square, a circle or a rectangle.

- **Size** – Size is simply the relationship of the area occupied by one shape to that of another.

- **Texture** – The surface quality of a shape – rough, smooth, soft, hard glossy, etc. Texture can be physical (tactile) or visual.

- **Colour** – Also called 'hue'. (We cover this in detail a little later in this section.)

- **Form** – This differs from shape in as much as it is three dimensional. So the shape may be a circle but the form is a sphere.

- **Balance** – As humans, we experience the need for balance in our everyday lives. We use it as we walk or run and need it to carry things. Balance is also necessary in other ways. We need to balance our time spent awake and sleeping, and our food intake and energy use. Balance is also important in products. A balanced design leaves the viewer feeling 'visually comfortable'. On the other hand, a product that is not balanced creates a sense of visual stress.

- **Proportion** – The word 'proportion' means one part in relation to another. Everyone has a sense of proportion concerning themselves, as compared to others. 'My nose is too long for my face'. 'She has long legs'. 'His eyes are too far apart'. All of these comments reinforce the idea that we see and have opinions about the relationship between one thing and another. Designers use their sense of proportion to make statements or to express a particular feeling about the way the elements of a product combine.

- **Symmetry** – Products can be symmetrical or asymmetrical. Generally, symmetry is pleasing to the eye. However, asymmetric shapes are also pleasing as long as the other elements of design such as balance and proportion are used.

Patent and copyright and consumer law

Patents, trademarks and copyrights

What is a patent?

A patent protects new inventions and covers how things work, what they do, how they do it, what they are made of and how they are made. It gives the owner the right to prevent others from making, using, importing or selling the invention without permission.

For an invention to be granted a patent it must be:

- new
- inventive in a way that is not obvious to someone with knowledge and experience in the subject
- capable of being made or used in some kind of industry.

It must *not* be:

- a scientific or mathematical discovery, theory or method
- a book, play, musical or artistic work
- a way of playing a game or doing business
- the presentation of information, or some computer programs
- an animal or plant variety

- a method of medical treatment or diagnosis
- against public decency or morality.

One of the best-known patented products in the UK is the 'cat's eye', the device that magically lights up from the middle of the road when you're driving along at night. The inventor of cat's eyes was Percy Shaw of Halifax, Yorkshire. When the tramlines were removed in the nearby town of Bradford, he realised that he'd been using the polished strips of steel to navigate. The name 'cat's eye' comes from Shaw's inspiration for the device: the 'eyeshine' reflecting from the eyes of a cat. In 1934, he patented his invention (Patent Nos. 436,290 and 457,536), and on 15 March 1935, he founded Reflecting Roadstuds, Limited in Halifax to manufacture the products.

Trademark rights may be used to prevent others from using a confusingly similar mark, but not to prevent others from making the same goods or from selling the same goods or services under a clearly different mark. Probably the best-known trademark in the world is Coca-Cola. The iconic shaped bottle and the word Coca-Cola are both registered trademarks. Any company that tried to use either the name or a bottle with the same shape would be in breach of the law. One of the best-known British trademarks is Cadbury. While Cadbury has a wide range of different products, the trademark is the common factor that links all the different products together.

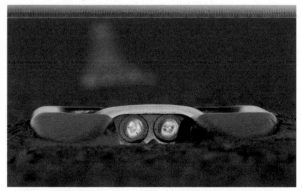

Figure 7.10 The cat's eye is a classic design that was patented, protecting the designer from illegal copies

Figure 7.11 The well-known Cadbury trademark

What is a trademark?

A trademark is a word, name, symbol or device that is used in trade with goods to indicate the source of the goods and to distinguish them from the goods of others. A service mark is the same as a trademark, except that it identifies and distinguishes the source of a service rather than a product.

What is a copyright?

Copyright is a form of protection provided to the authors of original works including writing, drama, music and art, both published and unpublished. The 1988 Copyright Act generally gives the owner of copyright the exclusive right to reproduce the copyrighted work, to prepare derivative works, to distribute copies of the copyrighted work, to perform the copyrighted work publicly, or to display the copyrighted work publicly.

Copyright law seeks to ensure that the rightful creator of the work is duly rewarded for their hard work. However, with the advent of the internet it has become increasing difficult to prevent copyright theft. Every time you photocopy from a book, download a music track without paying, or copy a CD you are committing a crime and stealing from the rightful owner.

Figure 7.12 Copyright is used to protect the author from illegal copying

Computer Aided Design (CAD) and Manufacturing (CAM)

Chapter 5 has already covered some aspects of CAD and CAM that you should encounter during your studies. However, the industrial implications, applications, advantages and disadvantages of CAD and CAM are also worthy of your consideration.

Have you ever wondered how the wings of the Airbus are designed and made in England and then flown to Germany for assembly on to the aircraft fuselage and how they fit perfectly first time, every time?

Using CAD, designers can quickly modify designs. Components associated with the modification can be adjusted automatically or

alerts can be issued to other members of the design team to show that a modification has been made.

The use of CAD also allows products to be designed in one country and manufactured in another. This offers companies the opportunity to both reduce environmental impact and improve efficiency by manufacturing close to where products will eventually be sold, minimising both monetary and environmental costs of transporting goods to market. Labour costs can also be reduced by remote manufacturing.

The use of automated production, which includes robotics, needs to be understood when designing. This is because the way an item is manufactured, the order of assembly and the shape of its components could affect the ability to use automated manufacture, which may have an impact on manufacturing costs and possibly the quality of the final product.

You must remember that when considering CAD and CAM you must reflect on industrial practice and not just your own experiences in school or college.

Marketing and economics

You already know the importance of the user or user group when designing. Designing products for the sake of designing is **not** product design. A solar-powered, left-handed, single-use toothpick is unlikely to have many people interested in buying it. No matter how hard you try, you are unlikely to sell many of them. Designing and manufacturing a product without knowing there is a real need for it would minimise the chances of ever making a profit from its design and manufacture.

Designing with a 'need in mind', which is related to a specific user or user group, **is** what product design is about. How to extend the potential market and the subsequent sales of the product can then follow. How companies identify a target market and subsequently a wider target market is part of what is called 'marketing'. They will use ranking and rating tests, consumer surveys, interviews and questionnaires to find out what people initially want and need and also what they think about existing products.

The results will inform initial design and development and also extend the markets for products as the product evolves. A single product can be designed and developed with different features for different markets based on the findings of the market research. The specification for different models of cars varies greatly depending on the country in which it is going to be sold. This is because each country has its own legislation and standards and the car manufactures need to take these into account.

Advertising is part of a marketing strategy and is used to attract more of a target group and persuade other people who might initially be in that target group that the product would be good for them.

Have you ever noticed that if there is a holiday programme on television then some of the adverts are for holidays? When there is a medical programme, some of the adverts are for medicines and cold cures? It is no coincidence; it is a deliberate ploy to target people on the edge of a particular target group.

Advertising is a very powerful and lucrative business. The positive and negative results of advertising can have both amazing and devastating impacts respectively.

There are many very successful advertising campaigns among them are Adidas, Nike and Coca- Cola, with which you will already be familiar.

When things go wrong it can be both expensive and embarrassing. American Airlines, who offer leather seating as a luxury to passengers, tagline was *'Fly in Leather'* but became *'Fly Naked'* in the Spanish translation of the company's advertising campaign. Did sales of flights increase?

The Advertising Standards Authority (ASA) is there to make sure all advertising, wherever it appears, meets the high standards laid down in advertising codes. Their website will tell you more about the rules for advertising, it also allows consumers complain online, and explain how the ASA is working to keep UK advertising standards as high as possible.

The section on globalisation of design and manufacturing is closely linked with marketing and economics as these two design influences are often working hand in hand together.

▶ Colour theory

Colour plays a vitally important role in the world in which we live. Colour can sway thinking, change actions and cause reactions. It can irritate or soothe your eyes, raise your blood pressure or suppress your appetite. If you put people in a red room it may make them angry and may induce them to overheat. On the other hand, put them in a blue room and they will feel cool and start to put on their jumpers.

Colour gives meaning to design. It informs our taste, our desires and – importantly for designers – our response to products. I like my toothpaste to be white with a bit of blue or red. Make it purple and I'll spit it out, regardless of the taste. Yellow cars sell for less than silver ones – it is estimated that a three-year-old yellow car will sell for anything up to £500–700 less than a similar silver one! A black bike looks heavier than a yellow one. And it's well known that wearing black can make people look slimmer.

As a powerful form of communication, colour is irreplaceable. We all learn that red means 'stop' and green means 'go'. Traffic lights send this universal message. Likewise, the colours used for a product, website, business card, or logo cause powerful reactions.

Colour theory

Colour theory encompasses a multitude of definitions, concepts and design application and we don't have the time or space to go into them all here. For your exam, you only need to know about some of the basic concepts.

The colour wheel

The basic colour wheel was first produced by Sir Isaac Newton in 1666, the year of the Great Fire of London. Whatever form the colour wheel takes, the pure colours (or hues) are built up from the primary colours red, yellow and blue.

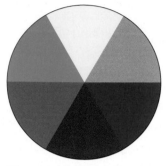

Figure 7.13 The colour wheel shows the relationship between colours

In traditional colour theory, there are the three primary colours, which cannot be mixed or formed by any combination of other colours. All other colours are derived from these three hues. In between the primary colours come the secondary colours, green, orange and purple, which are formed by mixing the primary colours together. Red and yellow makes orange, and so on. It becomes a little trickier when a primary is mixed with a secondary to make an intermediate or tertiary (third) colour.

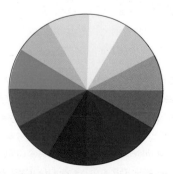

Figure 7.14 Colour wheel showing tertiary colours

Because a primary colour is mixed with a secondary to make the tertiary, the hues have a two-word name, such as blue-green, red-violet, and yellow-orange

Colour harmony

Harmony can be defined as a pleasing arrangement of parts, whether it is music, poetry or colour.

In visual terms, harmony is something that is pleasing to the eye. Harmony engages the viewer and it creates an inner sense of order, a balance, a sense of pleasure. When something is not harmonious, it is either boring or chaotic. At one extreme is a visual that is so bland that the viewer is not engaged. At the other is a visual that is so overdone and so chaotic that the viewer can't stand to look at it. Colour harmony delivers visual interest and a sense of order.

Figure 7.15 Harmonising colours can merge together to create vivid pictures

Take a look at the sunset in Figure 7.15 and you will see that the colours are very close to each other on the colour wheel. These are called 'harmonious' or 'analogous' colours. Analogous colours are any three colours that are side by side on a 12-part colour wheel, such as yellow-green, yellow, and yellow-

orange. Using analogous colours creates visual harmony. Alternatively, opposites attract. When it comes to colour theory this is certainly true.

Figure 7.16 Complementary colours on the colour wheel can create an impact

Complementary colours are any two colours that are directly opposite each other, such as red and green, or red-purple and yellow-green. In Figure 7.16 there are several variations of yellow-green in the leaves and several variations of red-purple in the water lily. These opposing colours create maximum contrast and maximum stability.

It is said that red and green should 'never be seen, unless there's something in-between.' Clearly, this doesn't apply in nature!

Figure 7.17 Red and green work well in nature

Nature provides a perfect departure point for colour harmony. In Figure 7.17, red, yellow and green create a harmonious design, regardless of whether this combination fits into a technical formula for colour harmony.

How colour behaves in relation to other colours and shapes is a complex area of colour theory.

ACTIVITY

Look at Figure 7.18 and compare the contrast effects of different colour backgrounds for the same red square.

Figure 7.18 Colours can be used to develop different effects when used with other colours to create contrast

Red appears more brilliant against a black background and somewhat duller against the white background. In contrast with orange, the red appears lifeless; in contrast with blue-green, it exhibits brilliance. Notice that the red square appears larger on black than on other background colours.

One other area where students often get confused is the difference between a shade, a tint and a hue. Put simply, shade and tint are terms that refer to a variation of a hue. Using a white or black pencil, for example, changes the shade of a hue by lightening or darkening it.

Figure 7.19 Shading blurs the distinction between two hues and should be used only with harmonising colours

Systems and structures

Structures

When designing structures we often examine the ways that natural structures are assembled and the way they behave under load. For example, internal doors use a honeycomb structure for strength and low weight. This is a direct copy of the structure of a honeycomb from a bee's nest.

Figure 7.20 The Atomium in Belgium. What is this structure based on?

Other structures can be as simple as the cross brace on a gate.

Monocoque is a construction technique that supports structural load by using an object's external skin. Monocoque construction was first widely used in aircraft in the 1930s. **Structural skin** is another term for the same concept.

Figure 7.21 Crossed braced bridge

ACTIVITY

What other product can you think of which are made using monocoque construction?

The way a pair of trousers is constructed and sewn together is based upon the designers understanding of the structural needs of the garment. They will understand areas of the design that require greatest strength and will ensure that the construction method to be used is suitable.

Figure 7.22 Structural details of Denim jeans

Systems

A system is a combination of components, which work together to perform an activity. Simple mechanical and electrical systems together with pneumatic and hydraulic systems control many products around us. When your waste is collected by the Council lorry a whole range of electrical and hydraulic systems are working seamlessly together. There is a braking system, a lighting system, a steering system, a heating system, a lifting system, a compressing system, a warning system, a sound system and possibly a communications system.

They all operate using three main stages:

INPUT → PROCESS → OUTPUT

- The input will vary between different systems. Inputs may include switches, mechanical movement, pressure sensors, temperature etc. Changes in the state of the input will be passed to the process part of a system.

- The process stage of the system will respond to the input signal and when pre-determined signal levels are reached the process stage will activate the output device.

- The output of a system could be the activation of an alarm signal or a lamp, the opening or closing of a valve, the removal of power, etc.

Example – Braking system

The driver puts his or her foot on the brake pedal and exerts pressure (INPUT)

The hydraulic fluid is pressurised which forces a piston to activate (PROCESS)

The brake pads squeeze on to the brake drums and slow the vehicle (OUTPUT).

▶ Energy

Some understanding of the sources and principles of different forms of energy is needed for any designer when designing quality products. Energy as a design influence is about the use of renewable and non-renewable energy sources and the various effects on the environment. It is also about the energy we get from food and bio-fuels, tidal, wind and wave power and the impacts they all have on the environment and on the design, manufacture and use of designed products.

Figure 7.23 Traditional 100 watt light bulbs to be phased out in favour of low-energy alternative. How will this affect the design of lighting products?

From January 2009 you can no longer purchase 100 watt light bulbs in the UK. Why and how this has come about is something for you to investigate to gain a better understanding of the issues. Whatever the reasons are, it is certainly going to affect the design of lighting products.

It will also affect the design of replacement products. The number of different low-energy light bulbs that can now be purchased, has grown rapidly.

Various sources of electricity can feed into the National Grid, which provides power for home and industrial use. However, not everything is powered by mains electricity. The use of batteries, rechargeable batteries and solar power is increasing. This brings with it disposal issues and design issues. If a doorbell is powered by solar energy then there will be a considerable difference in the appearance and location of the different components of the bell.

Batteries come in a wide range of shapes, sizes and power ratings. Some are suitable for some products and not suitable for others. The battery which powers a remote control for a television is considerably different in a number of respects to one which powers a hand-held food mixer.

KEY POINTS

When designing products there are always questions to ask and answer.

For example with batteries:

- What is the connection method?
- How will they be replaced?
- How are they to be held securely?
- What power rating might be needed?
- How will they be safely disposed of?
- Are rechargeable batteries better for the product?
- What size and shape is needed?
- Can the product be redesigned to house a specific battery?

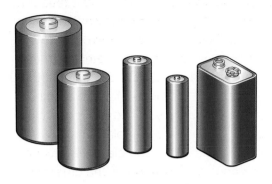

Figure 7.24 A range of batteries

◗ Scientific principles

Scientific principles sounds very technical but it is actually much simpler and easy to relate to than you might think.

You will need to understand how modern scientific principles and new materials have influenced the design of products.

We all know what a non-stick frying pan does. What we might not know is that it uses a relatively new material called Teflon®

manufactured by DuPont™. It coats the frying pan by a process called sintering. Sintering is a method of making or coating objects from powder by heating the powder up until the graduals stick to each other and, in the case of the frying pan, sticking to the pan itself.

You'll need to understand the properties of heat including transfer of heat and, of course, insulation. Think about radiators and loft insulation as a starting point.

The properties of materials such as weight, strength, resistance to distortion and the effects of heating and loading need to be considered. What happens to egg white when it is heated? How does PVA work? Simple basic qualities and properties of materials, components, ingredients and products are what you need to know.

How do simple levers work? How do the gears on a bicycle work? Why do we oil components? What is the effect of aerodynamic design of motor cars?

KEY POINTS

- Technically Teflon is poly(tetrafluoroethylene) or poly(tetrafluoroethene) (PTFE). It is a synthetic fluoropolymer which was accidentally invented when attempts were being made to create a new refrigerant. But you don't need to know all that! What you need to know is what it is, what it does and how it is used in products. It's a non-stick coating which can be applied to certain metals and ceramics! That's it!

Figure 7.25 One egg or two?

▶ Health and Safety

Health and Safety (H&S) is not only important when you are working in your design studio or workshop; it is also of vital importance when products are being designed and has to be taken into account during manufacture of products, but also when the products are being used. Indeed the safe disposal of products at the end of their useful lives is not only an environmental issue but is an H&S issue too.

Remember health and safety is more than just practical safety rules. It has wider implications and you need to understand the application of health and safety to products and their manufacture.

Figure 7.26 Fire alarm: Seven cabs burst into flames in London in the summer of 2008, including this one in the Haymarket

In September 2008 London Taxis International ordered an emergency recall of about 600 TX4 cabs with the backing of the Public Carriage Office (PCO). Why? Well, after seven taxis burst into flames in London that summer and at least 11 had also caught fire across the country urgent action was needed.

The 600 taxis were all returned to the manufacturer for investigation and repair.

The UK manufacturing sector comprises of diverse industries and employs 3.2 million workers. In 2007/8 there were 25,900 work-related injuries reported to the Health and Safety Executive (HSE). This included 5200 major injuries, of which 35 were fatalities.

The main causes of injury in the industry were:

- injury while handling, lifting or carrying
- slipping, tripping or falling on the same level
- hit by moving, flying or falling object and knife cuts
- skin disease, e.g. dermatitis
- respiratory disease, e.g. occupational asthma

Just how many of these injuries could have been avoided by changes in the design of the products being manufactured, how they were manufactured and what materials were used cannot be calculated. But certainly the design of products is a contributory factor.

▶ Globalisation of design and manufacturing

'Keep one eye on the imagination and one eye on the cash register.'

Raymond Lowey

Globalisation – what is it and why should we care?

The word globalisation (or internationalisation as it is sometimes called) refers to the increase of trade around the world, especially large companies producing goods in many different countries. Increasingly, companies are moving their operations, especially manufacturing, to a different country. For example, clothing manufacturers sell their goods in large high street shops in the UK, but the clothes may well have been manufactured in China. To some extent, selling goods that have been produced elsewhere has always happened, but during the last few years it has become commonplace.

When companies move their entire manufacturing operation overseas, it is known as '**off-shoring**'. For the manufacturer, the advantages of off-shoring are clear: lower costs, more profits and a ready supply of workers. The same reasons are often cited when companies such as banks and telecommunications organisations set up call centres in countries such as India, or when clothing manufacturers such as Nike get their trainers manufactured overseas. When companies do this, this is known as '**outsourcing**'. Outsourcing means that rather than a company doing everything themselves, they contract a second company (not necessarily overseas) to deliver a service or product for them.

CASE STUDY: DYSON

One of the best-known designers and entrepreneurs in the UK, James Dyson, moved his entire manufacturing plant from Wiltshire to Malaysia in 2003.

'Entrepreneur and euro enthusiast James Dyson was involved in a fresh row over exporting jobs yesterday after announcing he planned to switch production of washing machines from his base at Malmesbury, Wiltshire to Malaysia with the loss of 65 jobs.

The decision means the end of manufacturing for Dyson in Britain after last year's decision to move vacuum cleaner production to Malaysia, where production costs are 30pc lower. The transfer resulted in the loss of 800 jobs.

Unions reacted furiously to the washing machine announcement but there was a more restrained response from the Trade Department, where regret at the job losses was mixed with praise for Mr Dyson's contribution to innovation. Last year Tony Blair told MPs he was "deeply disappointed" at the Malaysian transfer.

Figure 7.27 James Dyson moved his entire manufacturing plant overseas

The joint general secretaries of Amicus – the engineering union – Roger Lyons and Derek Simpson, vied in their condemnation. Mr Lyons said Mr Dyson was like pop star Britney Spears singing "Oops I did it again" after last year's vacuum production decision. "He has no commitment to his workforce and is a desperately bad example to the rest of the sector." '

Daily Telegraph, 21 August 2003

The case study about Dyson sums up most of the issues that concern people about globalisation. While recognising the potential of increased profits for companies, many people are concerned that by moving to other countries, where the labour costs are much lower, foreign workers may be exploited. While the vast majority of companies have strict employment laws and treat their employees well, concerns continue about sweatshops, child labour, long hours and very low wages. An article that appeared in *The Independent* sums up some of these concerns:

In the crowded sweatshops of China's Pearl river delta, the world's toys are churned out, not by Santa's elves, but by 1.5 million peasant girls toiling through shifts of 12 or 14 hours, inhaling toxic fumes.

A 10-year campaign to introduce basic workers' rights has barely begun to improve the shabby treatment of the girls, new research shows.

'The Chinese toy factory workers are more exploited than before', said May Wong of the Asia Monitor Resource Centre who investigated the toy industry, with the Hong Kong Christian Industrial Committee. Another investigator, Monina Wong, author of a soon-to- be-published report for the Hong Kong Coalition for the Charter on the Safe Production of Toys, said: 'Wages have actually gone down, there is so much surplus labour. Conditions have improved a little, especially in overtime because big buyers are putting pressure on sub-contractors.'

But workers still have no contracts or unions, and little protection from owners who sometimes withhold part or even all of the wages due.
The Independent, 24 December 2002

▶ Sustainable technologies

Sustainability means the ability of something to last. Designing a product to last has become a particular concern in the 21st century. The growing amount of unnecessary waste and the impact of energy usage and carbon emissions on the environment have led to designers and manufacturers increasingly incorporating sustainability into their designs.

Reduce, reuse and recycle

The increase in the amount of waste as a direct result of excessive packaging and such things as throwaway wrappers, labels, and posters has led environmental campaigners to develop what they call a 'waste hierarchy'. A hierarchy can be thought of as a pyramid.

The largest part of the pyramid – the base – is the worst place to be on the hierarchy. The base represents disposal, that is, we carry on merely dumping or disposing all of our rubbish. As we move up the pyramid, we pass through stages such as recycle, reuse and reduce. Each of these stages in turn

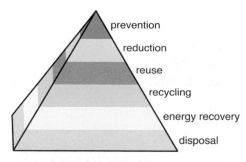

Figure 7.28 The three 'Rs' in the waste hierarchy are reduce, reuse and recycle

moves us higher up the pyramid towards the ultimate option, which is to prevent making the waste in the first place.

Do you know how much waste you make? It is estimated that every adult in the UK generates around one tonne of waste each year. This vast figure takes into account both the household and industrial waste created by things we buy or use.

Figure 7.29 We generate a lot of waste

Waste disposal is a major issue for the government and the community. If we minimise waste by reusing and recycling, we can cut waste by up to 50 per cent. Reducing waste also reduces litter. Remember, everything you drop on the street could find its way to beaches, roads, waterways, farmland and parks.

Reduce

Waste can be reduced by choosing products that can be used productively, recycled in our local area, and have minimal packaging. A good example of this is buying fruit in a loose form rather than in pre-packaged, shrink-wrapped plastic containers.

Reducing your waste also means saving resources. Too much energy is wasted in the

home. For example, lights left on, doors left open, electrical items left on standby or poor insulation – as much as 40 per cent of the energy used in homes is wasted.

Figure 7.30 A low-energy light bulb is a good example of a product designed to reduce energy usage

Reuse

Wherever possible, we should reuse containers and packaging. Increasingly, people prefer to use glass bottles rather than plastic because once they have been sterilised they can be refilled.

- Look for products in reusable, refillable or recyclable packaging when you shop.
- Donate unwanted clothing, furniture and white goods to charities.
- Enquire if goods can be repaired rather than replaced.
- Use rechargeable batteries rather than single-use batteries.
- Use glass bottles and jars, plastic bags, aluminium foil and take away food containers over and over again before recycling or disposing of them.
- Carry your lunch in a reusable container rather than disposable wrappings.
- Reuse envelopes and use both sides of paper.

Recycle

We can all reduce waste by recycling materials so they can be made into useable products. Paper, glass and plastic recycling are becoming commonplace. Another popular form of recycling is composting, the process by which food waste is broken down and used in gardening.

Figure 7.31 Recycling helps to reduce waste and reduce energy consumption

Recycling recovers materials used in the home or in industry for further uses. You should only recycle after you've tried to reduce and reuse.

Recycling has environmental, economic and social advantages:

- It generates pride and environmental awareness
- It helps prevent environmental pollution
- It saves natural resources
- It conserves raw materials used in industry
- Making products from recycled ingredients often uses much less energy than producing the same product from raw materials
- Recycling reduces the amount of material dumped in landfill sites and helps our waste disposal problems

- The life of goods is extended, which saves money

Recycled goods have already saved resources and raw materials and helped reduce the overall quantity of waste. Remember, 'recycled' means the product is made partly or wholly from recycled materials and 'recyclable' means the product is capable of being recycled. If you don't make an effort to buy recycled goods, you're not really recycling.

A good example of this is the paper pen – a ballpoint pen with the main body made from recycled paper.

Figure 7.32 Paper pens are a good example of a sustainable product

Sustainable design imperatives and design tools

Life-cycle analysis (LCA)

One key design tool now used when designing new products is life-cycle analysis (LCA) or what it is sometimes termed 'cradle to grave' (CtG) assessment. Life-cycle analysis goes from manufacture ('cradle') through the use phase to the disposal phase ('grave'). For example, trees produce paper, which is recycled into low-energy insulation, then used as an energy-saving device in the ceiling of a home for 40 years, saving 2,000 times the fossil-fuel energy used in its production. After 40 years the insulation is removed and replaced. The old fibres are disposed of, possibly incinerated. All inputs and outputs are considered for all the phases of the life cycle.

In broad terms, there are three stages to life-cycle analysis. These are:

- Stage 1 – goal and scope
- Stage 2 – inventory analysis
- Stage 3 – impact assessment.

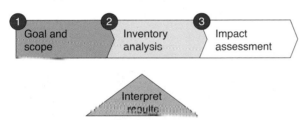

Figure 7.33 The three stages of LCA

Stage 1 – goal and scope

At this stage, the designer and manufacturer would consider what the final goal or aim for their product is. For example, a shoe manufacturer may want their shoes to be carbon neutral. Put simply, this means that the amount of carbon emitted during the life of the product is balanced by a carbon or energy saving during its use. For example one way of achieving this for airlines is to plant the equivalent number of trees needed to absorb the carbon emitted during their flights. This is called 'carbon neutralising' or 'off-setting'.

Stage 2 – inventory analysis

Once the goal has been established, the next stage in the LCA is the inventory analysis. An inventory is merely a list of all the systems and processes and events in the entire life of a product. Therefore, in simple terms, this means collecting all the data on the likely energy to be used at each stage.

Stage 3 –impact assessment

The final stage in LCA is the impact assessment. It is here that designers and manufacturers consider the impact of each of the processes that the product undergoes during its life. It is at this stage that decisions can be made about the various processes that will help to achieve the goal. For example, they may decide to take out one or two processes or to change a particular process or component to reduce the energy used.

▶ Economics of manufacturing

A number of the other design influences will automatically affect the economics (and scale) of manufacturing. For example sustainability and the need to reduce waste, globalisation of design and manufacturing the supply of raw material, labour and transportation costs, and social, moral and cultural issues where certain employees might not be able to work with leather but might be able to work with leatherette.

Others factors also affect the economics of manufacture and the associated design activity.

7.2 FASHION

Many people want to keep up with fashion. To be fashionable is to look 'cool', whereas failure to keep up with fashion can mean the opposite. Fashion dictates styles, colours and materials. Virtually everything we touch, wear, use or even live in is touched by fashion. Many of us associate fashion with models on a catwalk, but it has a lot more significance than that. So what is fashion? And who decides what the next fashion should be?

Fashion is hugely important to designers because failure to embrace fashion trends in their designs can lead to commercial failure.

Fashion and style in clothes is largely dictated by the large 'fashion houses' of Europe. These fashion houses employ the most famous clothes designers such as Karl Lagerfeld, Yves St Laurent and Stella McCartney. It is their vision, flair and creativity that influence the fabric designers, as well as the designers from the large high street stores such as Marks and Spencer and Next. No single person creates a fashion; it emerges from society and from trends both within the design world, and at large. For example, trends in music can lead to the start of a fashion trend.

CASE STUDY

During the 1970s, punk rock set out to offend, to challenge authority and to be a channel for venting the anger and frustration of young people. It was, in many ways, a reaction to the peace-loving generation that went before it. The style of clothing worn by bands such as the Sex Pistols quickly began to influence fashion designers. For example, ripped black fabric and leather, liberally covered with zips, became incorporated into mainstream clothing. Leading fashion designers such as Vivienne Westwood, long recognised as the mother of punk fashion, started to use new materials such as PVC in her designs. So while punk was a trend, elements of the punk style became fashionable. However, fashion by its very nature does not stay for long, and soon punk was replaced by a more feminine, extravagant style

known as 'New Romantic', a fashion that drew on style elements from the 18th century. To a large extent, fashion determines the way we dress.

Figure 7.34 Trends in popular style leads to fashion trends

7.3 OBSOLESCENCE

When it comes to other products such as cars, watches and MP3 players, we have to look beyond fashion to what has become known as 'obsolescence'. The term basically means to become obsolete – the point when a product is no longer of any use. However, people do not just replace their belongings when they no longer work. For example, people change their TV because they want the new model. Designers and manufacturers could, if they wanted, make products that last very much longer (possibly for ever). For a manufacturer, however, it is crucial to upgrade their products in order to create new markets and create demand from consumers. Obsolescence is therefore not just a process of a product wearing out naturally. It is a process that, either through technological development or through new design, leads to a product becoming prematurely 'out of date'. In design terms, obsolescence therefore has taken on a new meaning.

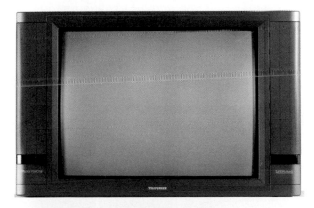

Figure 7.35 People rarely change their televisions just because they no longer work

▶ Technological obsolescence

Technological obsolescence happens when technological innovations or improvements supersede the existing technology. For example, the development of CD ROMs led to the demise of the audiotape. DVDs replaced videos, and so on. This kind of technological obsolescence can be seen throughout the world of product design. Almost as soon as a product is launched it becomes obsolete. This is because of the rapid development in technology and the need for designers and manufacturers to compete with their rivals in order to keep ahead.

As consumers, we either have to make do with the old model or buy the new, improved version. Huge demand for these new products as a result of clever marketing can put enormous psychological pressure on consumers. People also experience peer pressure, which has the effect of making them want to keep up with friends, family and colleagues. So technological obsolescence has a huge influence on the design of products.

Figure 7.36 New technological developments quickly make products obsolete

Planned obsolescence

Obsolescence, as described earlier, is not a new phenomenon. As far back as the 1930s the regular updating of products such as cars started what became known as 'planned obsolescence'. The term was first coined in the 1960s by the author Vance Packard in his famous book *The Hidden Persuaders*. Packard used the term to describe the process by which manufacturers deliberately changed the style of products on a regular basis, thus making the 'older' version appear out of date.

In reality, planned obsolescence was borne out of a genuine manufacturing need. In the car industry, heavy, automated machine tools had a life cycle of around three years. After this time, the tools began to wear to such an extent that they had to be replaced. It became obvious to manufacturers that rather than merely replace the old tools, they should update the car's design in order to boost customer interest. Thus, the typical three-year life cycle for cars started. This 'planned obsolescence' soon spread to all industries and as the pace of technological change accelerated, the three-year cycle became much shorter. In some cases, as with computer games, software and mobile telephones, the product life cycle can be as low as a matter of months.

Figure 7.37 A range of Nokia telephones from the year 2000 to the present

CASE STUDY

A mix of technological development, fashion and planned obsolescence makes mobile phones the 'must- have' fashion accessory of our age. Before the invention of the mobile phone, there was no such thing as texting. Now, every day, more than six billion text messages are sent. This exceeds the number of people who live on the planet. When you consider that the very first recorded text message was only sent on 3 December 1992 and said 'Merry Christmas', things have changed very rapidly since then!

7.4 NEW AND EMERGING TECHNOLOGIES

Rapid developments in technology mean that new materials are continually being produced. These new materials are changing the design of everyday products. From the development of enhanced food ingredients through to self-ironing fabrics, new materials are providing new and exciting opportunities for designers.

Composites

Many of the most impressive new materials are made by combining two or more materials together to form composites. Many of these composite materials have remarkable properties. A good example of a modern material is carbon fibre. Carbon fibre is a composite made from carbon fibres bonded with a flexible plastic resin. The resultant material is very lightweight yet incredibly strong. Such is the strength of this new material that it is used to form the driver's cab on Formula 1 racing cars, which can absorb the forces from collisions at speeds of more than 200 mph.

Modern materials such as carbon fibre are changing the way products are designed and made.

Smart materials

Many modern materials are now termed 'smart' because they respond to a stimulus such as electricity, heat, or light. The response they make may be to shrink, to change colour, to light up, or to become thinner or thicker. For example, oil is one type of smart material. When it heats up it gets less viscous (thinner). When it cools down it goes back to its original viscosity (thickness). A car will only operate at peak performance when the oil has reached the correct operating temperature. For racing cars this is a drawback, so engineers have designed a smart oil that changes its viscosity when electricity is passed through it. In this way the driver can control the viscosity, meaning

Figure 7.38 Modern Formula 1 cars use high performance materials that are lightweight and incredibly strong

that peak performance can be achieved even when the engine is cold, merely by passing an electrical current through it.

In the commercial manufacturing of products, many smart materials are now being incorporated into exciting products. A good example of a smart material is photochromic ink. These inks can change their colour, depending on the temperature. When used in a garment such as a T-shirt, the fabric changes colour as the person warms up. Although this is a novelty example, colour-changing pigments can have more serious uses. For example, when used in surgical cloths they can indicate a rise in temperature, a common sign of infection. Photochromic pigments can be added to any moulded, dyed or mixed material. Hence, colour-changing products are commonly used with plastics, food (such as cake icing) and textiles.

The pace of innovation and change

In 2006, Karl Fish, a US high school teacher, produced what has become one of the world's most widely read and reproduced PowerPoint® presentations. The presentation, eventually called 'Shift Happens', was placed on the internet and quickly spread around the world. The presentation highlighted the changes that information and communication technology is having on the world.

As students, you are living in a time of unprecedented change, a time where rapid technological development, especially in terms of our ability to communicate, has vastly improved. It is fascinating to know that this year there will be more new information produced than has ever existed before.

Karl Fish estimates that one week's worth of

information contained in the *New York Times* newspaper is more information than a person would have come across in their whole lifetime during the 18th century. Information has changed the way we think, the way we live and the way we learn. It is estimated that over three billion questions are asked on Google alone every month. Karl Fish asks the question, to whom did we ask the questions before Google? Of course, the answer to this question is that people used encyclopaedias and libraries. However, now that access to knowledge is instant, we ask a lot more questions.

Figure 7.39 Karl Fish states that over three billion questions are asked on Google every month

One of the main reasons for this huge change in information technology is the development of optical fibre technology. Optical fibres are very fine glass tubes that can transmit information in the form of light, at vast speeds. This technology is now used throughout the world to provide high-speed communication. For example, a single strand of third-generation fibre optic cable (about the thickness of one strand of your hair) can carry 10 trillion (10,000,000,000,000) bits of information per second. This is equivalent to 1,900 CDs or 150 million telephone calls every second.

This pace of development is predicted to triple every six months and to continue for a further 20 years. What this means for us as individuals, is that the way we learn and work will continuously change. The information that we learn today may well be out of date before we use it. For example, it is estimated that for students who follow a computing degree, the information that they learn in their first year will be out of date by the time they enter their third year. For people at work, it is clear that industries will continually change. Companies will have to change continuously to adapt to the changing technologies and needs of consumers.

Figure 7.40 Third-generation fibre optic cable can carry ten trillion bits of information per second

The development of new technology has changed the way that designers work. Traditionally, designers did everything by hand. They sketched their ideas and then worked these up into three-dimensional (3D) representations. Often, product designers had to make 3D models in order to test their ideas. They would then have to commission someone to make a prototype. At various stages throughout this process they would have to start their drawings again because of changes to the design (sometimes very minor changes).

As you can imagine, this took a long time and was hugely frustrating! Nowadays, however, designers can work directly from their own computer without leaving the comfort of their own home. Most 3D modelling has been replaced by Computer Aided Design (CAD) modelling and changes to the drawings can be made at the flick of a switch. Infinitely detailed models can be made by rapid prototyping. Furthermore, the designer can keep in constant contact with the client and manufacturer by email. Meetings can be virtual, using video conferencing (similar to using a webcam on Skype) and the delays caused by travelling and postage have been removed.

So, in simple terms, distance is no longer a barrier. Designers can live in one country, work for a company in another, and have their designs manufactured in a third. We live in a global community. This presents people with challenges, but also with exciting opportunities.

7.5 HOW HAVE THE DESIGN INFLUENCES BEEN APPLIED?

Designs which stand the test of time are often referred to as **classic designs**. These designs transcend fashion and encompass image, quality and innovation. The five famous classic designs detailed below all meet the requirements of a classic design. In design terms, each of the products has become an icon of design.

▶ The iPod (2001)

Arguably, the iPod represents the most popular and contemporary example of a design classic. Apple's iPod embodies the factors of simplicity, usability, and new technology that characterise revolutionary and highly popular objects. The iPod is destined to change, not because it is in any way flawed, but because of the very nature of the product and the rapidly emerging and developing technology that defines it.

However, the style of the iPod is the feature which sets it apart from so many other hi-tech devices we use. It doesn't have a mind-boggling array of controls, devices and functions – quite the reverse. The beauty of the iPod for many people lies in the way that such a complicated piece of technology can be made to appear so simple to the user. The use of white and chrome (symbols of purity and futurism) embody the forward-looking attitude of Apple's designers. Its sleekness is probably its greatest asset, that along with its functionality and smoothness of operation.

The iPod is not just a design classic: it is a symbol of an era. It marks a moment in time and in many ways defines a generation. We are a society for whom personal music, communication and media of all kinds are right in our back pocket.

▶ Pot Noodle (1977)

The Pot Noodle brand is considered to be the original instant hot snack. It is the UK's biggest-selling pot snack, with a huge 77 per cent market share. At the time of its launch by Golden Wonder in 1977, the UK was undergoing a marked lifestyle change. It coincided with many women deciding to go back to work rather than stay at home to look after their children. Inevitably, this led to less time for food preparation and home cooking.

Fuelling this new market came a whole range of convenience foods. This was not just a case of clever design. The introduction of these new convenience foods was due in large part to the development of new food production techniques that made such products possible. The Pot Noodle led the way, with its novelty value and simplicity. Never before had it been possible simply to boil a kettle in order to prepare a hot meal.

Figure 7.41 The iPod, designed by Jonathan Ive and the Apple design team

Figure 7.42 Pot Noodle is probably the best-known instant noodle product in the world

Figure 7.43 The Metropolitan Police whistle, designed by Joseph Hudson, Acme Whistles

At the time of its launch, convenience was the future and noodles were considered exotic. The original idea of 'cup noodles' came from Japan and has since captured the hearts and stomachs of our nation's youth. As with many of the world's top brands, the Pot Noodle has not stood still. A wide variety of flavours have been created and clever marketing campaigns have been developed to keep Pot Noodle ahead of its rivals.

Metropolitan Police Whistle (1883)

Not all iconic products naturally spring to mind when thinking of design classics. One such product is the Metropolitan Police Whistle. This simple yet ingenious product was invented in 1883 and is still in use all over the world today. Not only is the design iconic, but it functions so well that while new designs for whistles have come and gone, none to date have been able to improve on it. So, as with all iconic products, this whistle fulfils one of our key criterion for a design classic, namely that its design has stood the test of time.

The innovation in this enduring design, which is still in use by 120 national police forces

around the globe, is the sound. It can be heard from up to two miles away, originating from its iconic barrel shape. The design is small and therefore fits easily into the officer's shirt pocket. This dearly loved whistle has truly stood the test of time. In addition to its function, it has a clever feature that allows it to be held by an officer's teeth, so freeing up their arms if necessary. Many a criminal has been nabbed by both hands while simultaneously being deafened by the shrill blast of the trusty whistle.

When it comes to iconic products, the rule (as our best designers know) is, 'If it ain't broke, don't fix it.'

The Dr Martens boot (1960)

On 1 April 1960 the first pair of Dr Martens boots were made in a small factory in Northamptonshire. The manufacturers were Griggs and Co Ltd and the boots were named the famous 1460 to mark the date that they were made (1 April 1960).

A key feature of the footwear is its air-cushioned sole (dubbed 'bouncing soles'), developed by Dr. Klaus Martens, who was a doctor in the German army during the Second World War.

Figure 7.44 The classic Dr Martens 1460 boot

Central to the success of the Dr Martens brand has been the company's corporate image. The key features of the product, together with the use of the sun or globe, are encompassed within the successful logo. The clever use of distorted text arranged in the shape of a boot relates the logo to the product. The simple addition of lines to the word 'Martens' further enhances the logo and adds to the image of the boot. Using the two colours black and yellow creates contrast. This ensures that if photocopied the image will remain strong. The use of 3D text on 'Air Wair' creates impact and lifts it from the page, and the deliberate misspelling of the word 'wair' cleverly emphasises the bouncing air-filled soles. Dr Martens has been recognised as one of the world's top 100 brands, an image the company proudly cherishes.

IKEA

Ingvar Kamprad established IKEA in 1943 as a mail-order business. Fifteen years later he opened his first store in Sweden selling flat-packed furniture. IKEA is now an international retailing business operating in over 30 countries, with well in excess of 70,000 employees. Sales have grown steadily every year and are now estimated to be in excess of 10 billion euros.

His idea was simple – good quality furniture, well designed and beautifully made, at a cost that will suit most people. He famously said that 'people have very thin wallets. We should take care of their interests.' Furthermore, unlike his rival furniture retailers at the time, Kamprad decided that there was no point in a company spending money assembling the furniture it intends to sell when customers can do it for themselves.

The IKEA logo

The IKEA logo is simple and bold, and like other successful logos (such as those used by Shell and McDonalds) it is known all over the world.

As with most successful brands, the logo follows the acronym SECRET, as it:

- is **s**imple
- is **e**asy to understand
- uses **c**ontrasting **colour**s
- is **r**elated to the company
- is **e**nlargeable and reduceable
- and is **t**ransferable onto a range of different products.

The name IKEA comes from the founder's initials **I**ngvar **K**amprad, the first letter of **E**lmtaryd, his family farm, and **A**gunnaryd, the village in Southern Sweden where Kamprad was born. The logo closely identifies with Sweden because it uses the colours of the Swedish national flag.

Also, the boldness of the lettering and the contrasting colours (blue and yellow being two of the primary colours) make the logo really stand out. The logo has a modern feel and the use of geometric shapes in its construction alludes to the clean lines within the furniture. Like many effective logos, its colours can be interchanged without losing its visual impact.

Figure 7.45 The bold and simple IKEA logo

Figure 7.46 The Swedish flag

ACTIVITY

Consider each of the five products in turn and ask the key classic design questions:

- Has the product been successful?

- Has the product defined a generation or marked a particular period in time?

- Has the product stood the test of time?

- Is the product innovative?

- Has the design started a new trend or direction in design?

- Has the product been influential?

CHAPTER 8

UNIT A551: DEVELOPING AND APPLYING DESIGN SKILLS

8.1 UNIT A551 'CONTROLLED ASSESSMENT'

By the end of this section you should have developed a knowledge and understanding of:

- the structure of Unit A551 and what evidence is required for each assessment objective
- what controlled assessment is and how it will affect the work you undertake
- the types of design activities that are suitable for this unit
- strategies to ensure that you achieve the highest possible mark for the work you undertake.

Unit A551: Developing and Applying Design Skills is a controlled assessment unit in which you undertake a design task. This chapter explains exactly what you need to do within Unit A551. Examples of students' work will be used to show you what is required and what is not required. It will also show you different ways of presenting your own evidence.

The structure of Unit A551

There are two important things to know about this unit before you start work.

The first is that the unit is a 'Controlled Assessment' task. This means that you will undertake the work in your school or college under the supervision of your teacher. The work that *you* produce will be submitted as evidence to OCR for assessment. You cannot submit work that was done by someone else or work that you have copied from someone else. The work you submit should not be done outside your classroom.

The second is that Unit A551, Developing and Applying Design Skills, is broken down into what OCR calls 'Internal Assessment Objectives' (IAOs). These are a set of simple criteria by which all candidates entering an examination can be measured and compared. They are called 'internal' because you do them under the supervision of your teacher in school. Your job is to make sure you present evidence that meets these IAOs to the best of your ability.

There are three Internal Assessment Objectives in this unit. Below is a summary of what they cover.

IAO1 is about the problem, the user and brief. It involves:

- details of a problem to solve
- details of the user/user group affected by the problem
- some evidence of the problem and of the user
- a short design brief.

IAO2 is about research on the user and existing products and a specification. It involves:

- research with analysis and conclusions on existing products
- research with analysis and conclusions about the user and user requirements
- a detailed and justified product design specification.

IAO3 is about ideas, development, modelling, communication and CAD. It involves:

- generation of a range of annotated solutions to your problem, selection of the most appropriate solution, giving reasons for your choice and development of the solution to a final conclusion
- consideration of the function, ergonomics, aesthetics and other details during your designing
- relating your ideas and development to your product design specification
- use of a range of communication

techniques with clarity and confidence, which includes modelling to test the feasibility of your ideas

- use of CAD as a design tool during your designing.

The important thing to remember about this unit is: *You do not have to make what you design!*

 EXAMINER'S TIP

Don't waste time on grand titles that identify the IAOs within your folder. You do not actually have to make reference to them at all. What you have to do is provide evidence of your work that satisfies the requirements of each of them.

▶ IAO1: The problem, the user and the brief

The Product Design Toolbox gave lots of information about how to identify a problem and a user and how to write a brief. Now you are going to use all you have learned so far to do this for real.

 EXAMINER'S TIP

Remember, there are no marks to be gained for researching problems or users or for making the decision about the problem you are going to solve. But there are marks for getting straight to the point, i.e. explain the problem that *you* are going to solve!

The problem or situation could relate to someone you don't know, a friend or relative,

---

or a neighbour – or your teacher could 'give' you a starting point. Your teacher might point you in a particular direction that they know will be good for you, and give you a strong lead as to the problem and the user. It is then your job to explain about the problem and user and produce individual evidence to support your design work.

It could be that, because of the way you have been taught in your school or college, you are given a free choice of problem selection. Whichever way, you must be realistic – remember that you only have about 20 hours. Select something that interests you and that you can accomplish without the support of your teacher and your school or college. Whatever you start must be seen through to the end.

KEY POINT

- All the work you submit must be your own. This is an examination course, and at this stage what you are doing is showing the examiners what you can actually do. You can practise and get as much help as you need throughout the rest of the course, but for the 'Controlled Assessments' (Units 1 and 3) the work must be your own.

In the case of Figure 8.1, the designer has spotted a problem that involves her neighbour. This is an excellent example, and would gain top marks. There is a clear problem with people who suffer from arthritis, who represent a tangible user group and who could be used as 'experts' to provide information. Also, this situation has been personalised, with the focus on the neighbour. There is some basic observational

evidence to support the problem and user details and it is well summarised in the design brief. The problem and the designer's target outcome are clearly presented. It is presented on a single page.

Figure 8.1 We know what is going on and what the target is

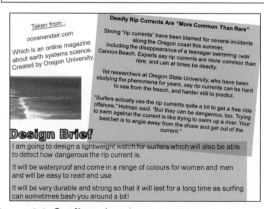

Figure 8.2 Surf's up!

Figure 8.2 shows another very good example of a problem that is also well presented to the examiner, but with not quite as much evidence provided of a user.

Figure 8.3 shows another excellent example of a designer identifying a clear problem for a specific user. There is evidence in abundance, with helpful data about the situation and a clear design brief that shows the intentions of the designer.

This is another excellent example that has a clear problem, a user and a brief. Also, the user has been extended to a user group, so it allows access to more expert witnesses as well as to the initial user. However, the user group is still a clearly defined 'closed' group of people and it will be easy to address their needs.

My client My chosen client is Peter Ward, who is a driving instructor for Ellies school of motoring. Peter ward has been an independent driving instructor since September 2006, before he was a driving instructor for BSM. Ellies school of motoring has an 80% pass rate, and they provide services for customers throughout Taunton, Bridgewater and the *Minehead* area including: Regular driving lessons, Theory test training, Hazard perception, Pass Plus, Motorway tuition, Refresher lessons, Advanced driving courses and Intensive courses. He uses his car as his office and brings about 9 items to work, some of these are; his Brief case, his glasses, his instructor books, driving props, a receipt book, pens, hearing aid batteries and chewing gum.

The Car is a FORD Fiesta Style, with a 1.4 liter engine. Its low fuel consumption makes it more environmentally friendly than most cars and its easy handling makes this the perfect car for teaching new drivers

pictures taken from Ellies school of motoring website (click here for more details)

Design need As he needs to carry so many things around with him, he would like me to design something which can store the documents and items of stationary he needs during the lesson, whilst the other books stay in his brief case. He would like it to fit into the boot of his car. He also has to show some clients documents. These documents needs to be kept safe whilst driving but be easily assessable. He needs something which can go in the back of the car whilst he travels to the location of the lesson, and be easily assessable when he gets there. Below is just some of the things he needs to travel with.

Range of users My product will mainly be used by driving instructors and their clients, but it may be used by any other person who uses their car as an office, such as; Taxi drivers and policemen. I have made a questionnaire about Ellies school of motoring and other driving schools, to find out how many people are involved with driving instructor and their cars. First of all I asked Peter ward some questions (red) then I did some research about Driving instructors (blue)

1. On average, how many hours do you work a day? 8 – 10 hours

2. on average, how many new customers do you get a week? 2-3 new customers

3. How many items do you bring to work? about 8

4. what is your most popular driving course? Standard beginners course.

5. how many driving instructors in the area (Somerset)? About 22

6. how many driving instructors in the UK? About 1,290,000

I asked the five that I asked Peter to a number of different driving instructors in the UK and Somerset by email. Click here to see my results.

My results show me that most driving instructor in this area:
• they work for about 6 to 8 hours a day
• they get about 1 to 2 new customers a week
• most of them bring about 7 items to work
• the most popular course is a standard beginners course

Design brief I am going to design and make something which fixes to all cars, which is strong enough to be used as a desk, and which can transport many items used by a driving instructor during a lesson. It needs to be able to be stored in the boot of the car and be able to fix to the front dashboard in comfortable position for the user.

Figure 8.3 Ellie's school of motoring

KEY POINT

- Have you noticed that all of these examples are solving a problem for someone else and not the designer themselves? They also have very clearly defined users who can be questioned as experts and will give excellent primary research results.

EXAMINER'S TIP

It is of vital importance that you have a clearly identified and documented problem and user before you move onto the next stage (IAO2).

Table 8.1 shows what you need to do.

Internal assessment objective	Total marks available	Approximate time it should take to do in hours	
1	6	1	
What you have to do is:		Total marks available	Approximate time it should take to do
• Identify a specific problem or situation • Provide details or evidence to support the problem			30 minutes
• Identify a user, users or user group • Provide details or evidence about the user			25 minutes
• Produce a short design brief explaining what you are going to do			5 minutes
Totals		6	1 hour

Table 8.1 **Mark summary IAO1**

▶ IAO2: Research on the user and existing products and a specification

We will look at aspects of IAO2 in several sections.

Research with analysis and conclusions of existing products

This part of the design process is to find out some data that will help to inform your own designing. If, for example, similar products to the one you are designing always contain a certain ingredient, or are always a similar colour, there must be a good reason for it. You will need to find out why, and then decide whether you need to include the same ingredient or use similar colours in your design. If you already know the reasons, then you can make 'informed decisions' with your own designing. For example, it would be a disaster if you designed a new set of traffic

lights that didn't use red, amber and green! The colours are in the positions they are and they are the colours they are for good reasons.

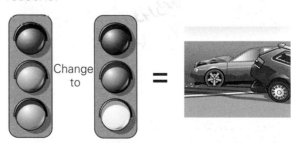

Figure 8.4 **Who designed those lights?**

A question you might ask is 'What information do I need to find out, then?' This is a very difficult question to answer because it just depends on what your problem is and what you are attempting to design.

Figure 8.5 shows some of the research undertaken by a designer who is designing a product for an 'electronic handbag' for a bride to hold information and data about her many wedding guests. We can gain some helpful information from the analysis, but also you will spot that some observations have limited value, for example the price tag of £52.

Similarly, with Figure 8.6, where the designer is designing a jar opener, there are some really useful bits of information to help with the designing and also some that are not. This is another example where time is being wasted on 'costs'.

Figure 8.5 Existing product research linked to an electronic handbag

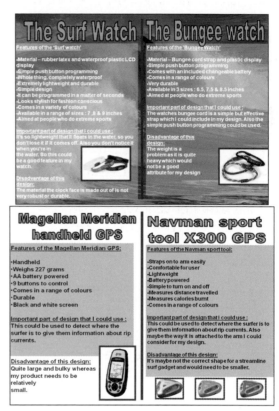

Figure 8.7 Research into aspects of the problem

This designer has chosen a topic that is of great interest to her, which has plenty of research and experts at hand and also has plenty of scope for creative and challenging design.

Her design brief says:

'I am going to design a lightweight watch for surfers that will also be able to detect how dangerous the rip current is. It will be waterproof and come in a range of colours for women and men and will be easy to read

Figure 8.6 Existing jar openers

and use. It will be very durable and strong, so that it will last for a long time, as surfing can sometimes bash you around a bit!'

Figure 8.7 shows some of her research into aspects of the problem. Notice the background in the first portfolio sheet, which can be produced within an electronic portfolio. The second portfolio sheet has had the background removed. It is easier to read but is not quite as exciting to look at.

EXAMINER'S TIP

As long as your work is easy for you and other people to read, then an appropriate background can sometimes set the scene for the mood of your work. Take care not to make it too complicated and also not to waste time on this, though.

Figure 8.8 is laid out in a different way, but has really useful headings that indicate that the designer has carried out good research.

The problem that this designer is trying to solve involves fixing something inside a motor vehicle. Table 8.2 shows the headings that are used. These are specific to the problem being solved and also specific to the products, and have a strong chance of getting good information to help the designing.

Figure 8.9 **Research into model aeroplanes**

The first portfolio sheet in Figure 8.9 has hardly any information apart from the materials used. The second sheet is much better, with more relevant information being gathered.

Figure 8.8 **Quality research**

Product	Purpose	How does it fasten to the car?	Good/bad points/ improvements for the product	Would the fixture method work for me?

Table 8.2 **Helpful headings**

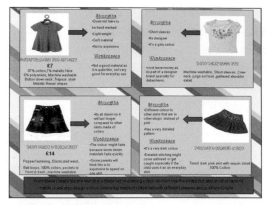

Figure 8.10 Children's clothes

This designer is designing clothes for three to four-year-old girls.

QUESTION 1

Look at Figure 8.10 and list the useful design information that you think is being gathered from this one design sheet.

Research with analysis and conclusions about the user and user requirements

In Chapter 2 of the Product Design Toolbox, we learnt how important the user is in the design process. This part of the research looks at information gathered about the user

and their particular needs, and also at other information that is not specifically based on *product* analysis.

First, let's look at some user information.

Figure 8.11 shows one sheet and not a lot of information gathered, but what there is links together nicely and has valuable ergonomic data. The measurement taken of the jar lids is good product analysis and the actual measuring of the hands is all about the user and provides very appropriate anthropometric data.

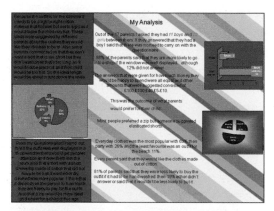

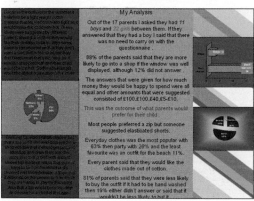

Figure 8.12 Same slide, different layout

Research - Ergonomics
Lid sizes and hand sizes

For this piece of research I used the jars in my kitchen to find out the average diameter of jar lids. I needed to do this to find out what size lids my jar opener will be used on and how big to make it. After measuring the lids of many jars from a wide range of brands I found out the that lid sizes range between 6 and 7.5 cm. Because of this I will design my product so that it can be used on any of these jar lid sizes. I also had to measure hand sizes as my product is going to have some sort of handle. I measured the width of hand palms and have decided to make the handle 15 cm long after finding 9 cm to be the largest I measured. This way it will always be long enough and could provide more leverage.

Figure 8.11 Remember the jar opener?

KEY POINT

- Make sure all of your work is legible and easily understandable.

QUESTION 2

Look at Figure 8.12 and see if you can work out why one sheet is easier to read than the other.

Figure 8.12 shows the results of a questionnaire used to carry out research with the parents of three to four-year-old children. There are written comments and also some simple pie and bar charts to illustrate the designer's research findings.

KEY POINT

- Questionnaires are a very good way of getting primary research but great care has to be taken to ask the right sort of questions. To go back to the example of the traffic lights on page 115, 'What is your favourite colour' is clearly not a useful question for a designer to ask!

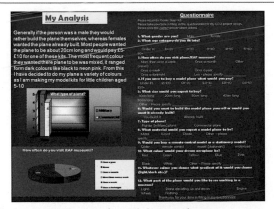

Figure 8.13 Users of model aeroplanes

A similar approach is used here, with both the questionnaire and results presented as evidence. The question 'What part of a plane would you like to see working in a museum?'

is an open-ended question. Open-ended questions are generally not a good idea to use on questionnaires, but, because there is a limit to the 'moving' aspects of an aeroplane, then in this case it should result in useful information.

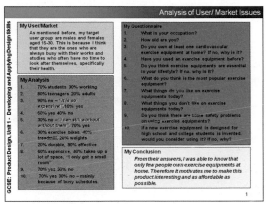

Figure 8.14 Alternative layouts and presentation of a questionnaire

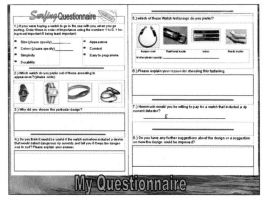

Figure 8.15 Quality of questionnaires

Figure 8.15 is an excellent example of a questionnaire that is well laid out and nicely presented. It is clear that the designer has put a lot of effort into it. It sets out a professional approach and it is likely that the users will respond positively to the questions in the same way. If you present a questionnaire on a scrappy bit of paper, you are likely to get scrappy responses that will probably not help you with your designing.

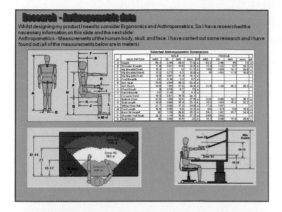

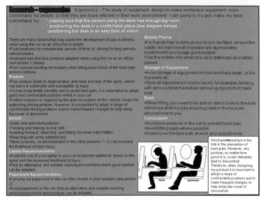

Figure 8.16 Appropriate and very detailed data

Figure 8.16 shows excellent examples of a mixture of primary and secondary research that is totally relevant to the project of an in-car, workstation storage unit.

Other research

In addition to investigating existing products and user requirements, quite often you will need some information that doesn't quite fit into either of these two categories.

The designer we met earlier who was designing a 'watch for surfers that will also be able to detect how dangerous the rip current is' looked into the formation of the waves and rip tides to inform her design

KEY POINTS

1. Questionnaires are not the only way to get information from a user or other expert. Face-to-face and telephone interviews, email enquiries, letters, observations and technical publications could all result in important information which will help you with your designing.

2. Using information from secondary sources, i.e. from published material, needs to be undertaken carefully. You could be wasting your time if you get no valid data, and then you will gain no credit for it at all. If you do secondary research, it must be really focused.

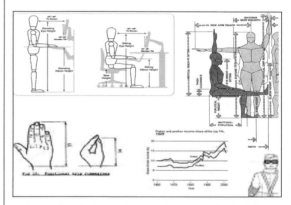

Figure 8.17 Generic anthropometric data sheets

3. There is only very limited value in using internet or textbook data on anthropometrics and it is unlikely to gain any marks. You will probably need very specific data. For example, if you are designing a bracelet, then the data you should collect is the wrist measurements of individuals within your user group.

work. Figure 8.18 would gain marks in the second strand of IAO2, which is concerned with the user and their requirements.

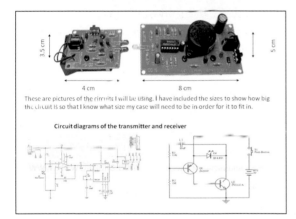

Figure 8.18 Rip tides

Figure 8.19 Electronic components

Figure 8.19 shows an example of components and circuits that the designer has researched and will incorporate into their design work.

A good designer will take legislation into account at some point. It is not always necessary for a student to do so, but if it is, then your research needs to be relevant and specific to the problem at hand. Figure 8.20 is an excellent example of this.

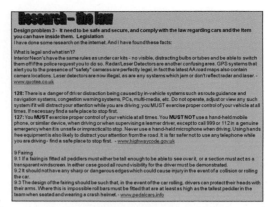

Figure 8.20 Researching relevant legislation

KEY POINT

- Don't copy irrelevant data into your portfolio and don't waste time explaining what ergonomics, anthropometrics, consumer law, copyright and patents are. What the examiners want to know is how these things affect your design ideas.

These buildings in Figure 8.21 could well be inspirational research on which to base shapes for a lighting problem, or even a storage problem. As they stand, the student would gain no marks for them, but with suitable annotation they are just what product design is all about – solving problems by thinking in unusual ways.

KEY POINT

- Mood boards without annotations won't gain you any marks. It is annotation explaining how the images might influence and support your designing that will gain the rewards. (Look again at Figure 8.21.)

Figure 8.21 Mood board – inspirational shapes for a lighting project

A detailed and justified product design specification

There is no set way to present your product specification – one way might suit you better than another. The important factors are: being realistic with the points you are going to address, remembering to include the user's requirements and then justifying each point.

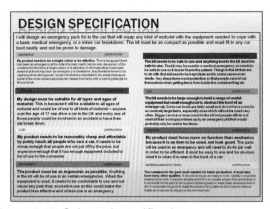

Figure 8.22 Colour-coded specification

Figure 8.23 Coloured specification

In Figure 8.22, the specification points have been given headers or titles to give structure to the product specification. This is not compulsory, but could suit your style of working.

Figure 8.23 has also used the technique of colour coding for each of the points. This can be very useful when referencing your

EXAMINER'S TIP

If you use headings, you must take care that they do not restrict you. If you use a random 'set' of headings you may not cover all of the points you need to, and so you will miss an opportunity and limit your designing.

Figure 8.24 Make sure your work is totally legible

KEY POINT

- Remember the point made about 'cost' and the Product Design Toolbox in Chapter 2? It is really best avoided unless you really tackle it in depth and this could be very time consuming to make it a valid exercise.

specification in IAO3. In Figure 8.23, you will notice that the designer references their work against their specification as they go along. Doing this will gain marks.

Keywords	Details
Function	The purpose of the product I am designing is to allow the brides to identify the guests that will be visiting their wedding without having to embarrass themselves in front of the guests you are talking to. It will tell the brides the personal details of the guests they have invited to their wedding.
Performance	The product I am designing can be placed in a handbag which she will be carrying with her or she will be wearing on the day she gets married. The electronic device will match with the dress she will be wearing on their wedding day
Size	The size of the product should be big enough for the bride to see information from it and it should be well discrete which the bride will be holding and if the guests did see it they will be quite offensive about how the bride doesn't know who the guests are
Weight	It matters a lot if my product is light or heavy. This is because that product will be held or worn by the bride all day long on her wedding day so it should be light as possible.
Target market	The group of people I am aiming my product is at the people who are getting married at the age of 20 – 40. The product I am designing creates an image of style, fashionable and wealthy
Life in service	This design is expected to last for about 2-3 years because mostly everyone who is getting married today and over the next year, forget their 3rd or 4th cousins, family or friend and the material I am going to use will last a long time and could use it afterwards to upload different information.
Aesthetics	The colour, the texture and the pattern will depend on the dress they will be wearing. The shape will depend on the size of the electronic device. This should make the product same as the dress they will be wearing and make it more trendy.

Figure 8.25 Using a different typeface can work wonders

Figure 8.25 shows how a different font can make your work more legible – but be careful, it could spoil the mood of your design folio. You will have to seek a good balance between the two.

Keyword	Answers
Ergonomics	I will make sure that the electronic device is easy to hold or worn for quite a long time and make sure that it is not that heavy. I also have to make sure that it looks like a product made by a professional – not that big or not that small and it looks really stylish and cool.
Materials	I would have to make sure that the product is made out of a material that is waterproof so that the electronic device doesn't get wet and blows up and that it is quite strong, cheap (so that they can buy it really easily).
Safety	I have to make sure that there are no sharp bits on the electronic device because they could hurt their hand whilst using the electronic device.
Cost	The product should cost at a low price in order to allow the consumers to buy the product I am designing really easily and that more people buy it.
reliability	Before my product is sold, it will be tested to find out the different problems that are held in this product.
Manufacturing/quantity	This product will be one-batch product because it should match the theme of the wedding dress. The electronic device will be made in factories and it should be able to match the dress that will be worn on the wedding day.
Possible conflict	This should be quite trendy, cost less and should match with the wedding dress the bride will be wearing. This is to make sure that the consumer can buy it easily, and the product is attractive so that the consumer will buy them.

Figure 8.26 Product specification

The product specification shown in Figures 8.25 and 8.26 has many really good specification points that the designer has justified and so is very informative. It guides the design process well.

QUESTION 3

Take note of the comments in Figure 8.26 against the heading of cost. Is this point helpful?

Figures 8.27 to 8.30 are laid out in a variety of ways. Some use headings, others bullet points or set the information out in a table. What they all have in common are specification points that will inform the designing, and justification of these points.

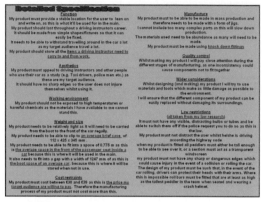

Figure 8.27

DESIGN SPECIFICATION		
Essential	**Desirable**	**Optional**
My product must be small and lightweight to be transportable for elderly people. Max size: 15cm long approx. Max weight: 1.0kg approx.	This product could be used by people who don't have a disability, or who just find fiddly keys annoying. (Non-exclusive)	Type of grip (e.g. T-Bar, Pistol). May be adjustable to suit particular needs.
Must use a colour that is attractive to the buyer, but also durable and practical for everyday use. (e.g. neutrals but with a flash of colour).	Should not be too bulky for user to cope with, or take up too much room in a bag. (Ergonomics).	Could have a return address (not the users) to allow it to be safely returned if lost.
Must hold at least three keys, (of various makes and sizes) and have an extra function (penknife etc) to add interest to the user/buyer.	The attachment to clip onto bag or accessory to be able to fold away when not in use, to prevent accidents or injury to user.	My product could be a sponsored item with a logo etc. relating to a company. This would reduce overall costs.
Must have an attachment to clip onto a bag or other user accessory, taking into consideration need for ease of access to their location.	Materials should be cheap but a give a high quality finish, and be robust enough to last for up to 5 years.	Could have a locating whistle or Led light to help if it ever gets lost.
Must be safe to use (no small dangling parts) and should have rounded edges where possible.	Should meet BSI standards relating to products designed by elderly users.	May be magnetic so that keys can attach more securely, or to allow for easy storage when not in use.
Must fit the average hand span of an elderly woman/man or disabled child (Anthropometrics). E.g. no larger than 24cm and no smaller than 10cm.	My Product should be produced in an environmentally positive way, using low impact machinery and renewable materials where possible.	Come in a range of colours to suit users tastes- allowing for some degree of personalisation.

Figure 8.28

Specification

- Must be easy to program because the surfers wants to be able to get into the water quickly and not spend all their time programming the watch.
- Must be aimed at a target market of teenagers who surf for fun to older people of around 30 or 40 years of age who surf for fun or surf professionally.
- Must come in a range of colours to appeal to a wide range of people. Also must either come in a range of sizes to fit a wide range of surfers wrists as all builds and age of people surf. Or it should be made out of a flexible material that will fit everyone.
- Must be aesthetically pleasing and stylish so the surfer will not be ashamed to wear or use it, and must be safe to wear and use in its surroundings.
- Must be very robust and durable so it lasts for along time without breaking, and be as near to maintenance free as possible.
- It needs to be battery run and the batteries need to be easy to change, or use rechargeable batteries so it will keep working without the power fading.
- Must be reasonably small (the face no larger than 4cm by 4cm) and lightweight (no heavier than 30g) so it doesn't get in the surfers way when they're surfing or catch onto their clothing as it could become dangerous.
- Must be made out of a water resistant material, for the clock face it should be made out of a robust plastic. For the strap a water resistant, shatterproof, comfy material that has a smooth and sleek finish to it.
- Must efficiently, effectively and simply display the time and danger of the rip currents so the surfer can easily read the information without hassle.
- It must be a simple and comfortable design that the surfer must feel at ease wearing for frequent use.
- Needs to be produced to be sold at £10 to £50 to be able to be bought by a range of wealth surfers.
- Must be produced so that it comes in a minimal amount of packaging and the packaging must be eco friendly. Must come with a small instruction booklet to instruct the user of how to operate the watch and its functions.

Figure 8.29

Specification

Size: Utilizing my survey questionnaire and my individual designs, I know that the size of my product will be no larger than the actual MP3. The MP3 player should at least have a length of 140mm and a width of 70mm. This would provide a range of MP3's enough space to fit inside the case.

Function: The function of my product will be to store MP3 players for suitable environments so that the users are satisfied with the product. The storage device should hold MP3's of various different shapes and sizes. This would increase the number of users that would buy the product.

Target user: The target users for my product are people aged from 15-35. This age range will effect my design and the purpose of the product. The design will have to be something unique and in demand for everyday use.

Timescale: The timescale for the production of my product is 20 hours. This will affect the complexity of the merchandise. The 20 hours will include folder work, practical work and design drawings. The folder work will consist of several different problems faced by existing products and how this will improve the product furthermore. The practical work will include the production of the model as well as the production of the main produce; and finally the design drawings will show a diversity of designs and will show a variety.

Environment: I intend to use the product in conditions suitable for my target users. These will include work places such as Offices, schools etc. The product will be able to sustain any problems that threatens to damage an MP3. The secure storage will adapt to its environment so it can protect the MP3 player.

Ergonomics: The product will be comfortable to use and hold. I will make it hand held so that the user is able to hold it with ease. In addition, I must consider the weight of my product as the users would find it difficult to carry heavyweight materials. These ergonomics are necessary if I was to make my product successful.

Aesthetics: The aesthetics will have to be smooth, sleek and modern. The unusual design will encourage the interest of buyers. The atypical product must contain essential features to exert a pull on my target users. These include the product being multipurpose as well as containing the product requirements. (Protect, hold and transport)

Figure 8.30

KEY POINT

- In the specification examples presented, not every specification point is informative and justified. But, if you look at them closely you will be able to identify some that are really good and some that are slightly less useful.

You can do this with any existing specification, whether it is one of your own,

ACTIVITY 1

(a) Read the product specification shown in Figure 8.30.

(b) Then take each point and rewrite it:
- with a good clear justification (to get higher marks)
- with the user in mind (because the user is so important when designing)
- using fewer words (to make it more concise and to the point)
- removing any 'useless' information (to make it clearer to follow)
- adding any further points you can think of (to make the specification more comprehensive).

one that has been given to you or one from this book. It is excellent experience and will improve your specification writing skills.

Table 8.3 shows what you need to do.

Internal assessment objective	Total marks available	Approximate time it should take to do in hours
2	23	4

What you have to do is:	Total marks available	Approximate time it should take to do
• Make a detailed examination of other similar products • Provide a detailed analysis of other relevant research	8	2½ hours
• Identify and collect all significant data about the relevant user(s) of product • Examine and collect other key data	7	1 hour
• Produce a detailed and justified specification which fully considers the user	8	½ hour
Totals	23	4 hours

Table 8.3 Mark summary IAO2

▶ IAO3: Ideas, development, modelling, communication and CAD

IAO3 has a number of strands that are different, but interdependent. Let's remind ourselves of the content that you have to produce in IAO3. Here is what each strand requires you to do.

For Strand 1, you need to have: generated a range of annotated solutions to your problem, selected the most appropriate solution, giving reasons for your choice, and developed the solution to a final conclusion.   **[25 marks]**

For Strand 2, you need to have: considered the function, ergonomics, aesthetics and other details during your designing. **[8 marks]**

For Strand 3, you need to have: related your ideas and development to your product design specification.   **[10 marks]**

For Strand 4, you need to have: used a range of communication techniques with clarity and confidence, which includes modelling to test the feasibility of your ideas.   **[8 marks]**

For Strand 5, you need to have: used CAD as a design tool during your designing.

  [10 marks]

All of the marks you gain in IAO3 will be from information and detail about your thinking when solving the problem you originally identified in IAO1, and related to the specification that you wrote in IAO2.

In the next section, we will look at a number of separate pieces of work and identify where they will gain marks in this section, strand by strand where possible, and at the reasons

KEY POINTS

- There are 61 marks for this section of the portfolio and the mark breakdown is indicated by the marks shown in the squared brackets [ ].
- The work in IAO3 must relate to your work in IAO1 and IAO2. You cannot gain marks in IAO3 if your designs do not relate to your design brief and product specification.
- Marks for communication and the use of ICT, Strands 4 and 5, can only be gained for work in this IAO. Work in IAO1 and IAO2 does not gain credit for these two skills.

why they won't gain them in other sections of this IAO.

What you need to do

You will need to come up with a number of different ideas to solve your problem. Each idea may not meet all your specification points and some ideas may just simply not work properly. That's fine, as long as you are continually thinking about the problem and the user.

As you sketch and draw your ideas you must also make reference to your design specification using annotations and also make detailed comments on things like the function, ergonomics and aesthetics of your ideas.

You will then choose an idea (or possibly parts of different ideas and combine them) and develop it further to address your written specification, as far as you possibly can. You may have to modify your thinking in the light of your design work, and to suggest that your

KEY POINT

- Things to avoid include simply stating things like 'this idea is aesthetically pleasing' or 'the handle is good ergonomically'. You must give detailed reasons *why* you think it is aesthetically pleasing or ergonomically good.

specification needs improving – and that's fine too. Just remember to record everything in your portfolio because the examiner cannot read your mind.

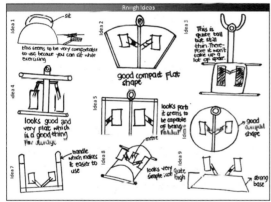

Figure 8.31 **A range of simply presented ideas**

Figure 8.31 shows a good range of simply presented ideas and will gain credit in Strand 1. The designer is solving the problem of emergency storage in a car. The comments at this stage do not really help us to understand anything else about the ideas, and so are unlikely to be of much value.

The drawings are all 'flat views' and have no 3D effects, nor has colour been used, but credit will be given in Strand 4 at a basic level for the evidence provided.

So, this represents a good start with ideas

and will gain some basic credit for communication skills. This is quite often the way designers start. They then go on to give more details as their thinking develops. At this stage, no marks would be awarded in Strands 2, 3 or 5 for this work.

Figure 8.32 **Simple ideas with some comments**

Figure 8.32 shows a similar range of ideas to that shown in Figure 8.31 but they are better drawn, using 3D. This designer is also looking at storage within a car. In fact, this designer produced even more different ideas than those shown, using this same format, and so gained really good marks in Strand 1 because he showed a really good range of thinking. The designer has made some comments, so marks in Strand 2 and possibly in Strand 3, if the specification has been considered, will then be awarded. There is some very limited credit given for the use of ICT in Strand 5. Word processing has been used and images have been embedded in a table.

KEY POINTS

- In Strand 5, there are up to two marks available for your ICT work presented being just word or data processed or simple (ICT) drawing.

- This designer has also made comments about their ideas with relation to innovation and flair. This is *not* required. The examiner will make final judgements on this matter and your time is better spent on other things, for example producing another design idea.

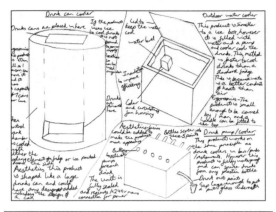

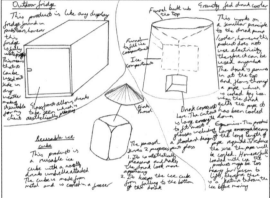

Figure 8.34 More ideas

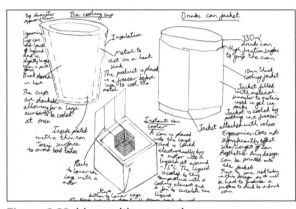

Figure 8.33 Ideas with annotation

Figure 8.33 shows three ideas for a drinks cooler. The drawings explain the ideas fairly well, but the annotation is very good and therefore addresses Strands 2 and 3 well.

The three ideas will gain marks in Strand 1 because they are different, but it is not a particularly wide range and therefore will not gain the highest marks available.

When you put Figure 8.33 to Figure 8.34, the range of different ideas is now quite extensive and so will get very high marks in Strand 1 and, because the annotations are about different points, the marks in Strands 2 and 3 will grow significantly. However, there is no improvement in Strand 4, the communication marks, as there is 'just more of the same style' of drawings. You will be rewarded for a *range*, not for quantity. There is no use of ICT, so no marks in Strand 5.

Figure 8.35 shows the same designer's work at the final ideas stage (the development work has not been shown here). The example

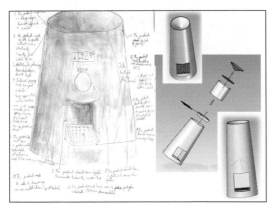

Figure 8.35 Final idea

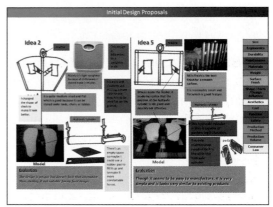

Figure 8.36 Development

shows CAD drawings of the final solution for this problem. There is a further improvement in Strands 2 and 3 because of the extra details provided, a slight improvement in Strand 4 as the drawing is of a slightly higher quality, and reasonable marks will be gained in Strand 5 for the use of ICT.

KEY POINT

- No matter how good your CAD drawings are, you can only gain up to seven of the ten marks available in Strand 5 if you have just drawn out what you have designed. To gain the full marks you must use CAD as a design tool during your actual designing.

Figure 8.36 shows some development in the *simple ideas* work you looked at in Figure 8.31. Marks will be increased in Strand 1 as an extension of the designer's thinking is evidenced. It also shows the use of a colour-coded specification, so there are good marks being gained in Strand 3 as the colour coding clearly shows how the ideas relate to the

specification. Marks are also being gained in Strand 2 as the function and aesthetics are being addressed too.

The communication has improved very slightly with the use of 3D, but there is also some modelling of both of these ideas. The modelling will be rewarded in Strand 4.

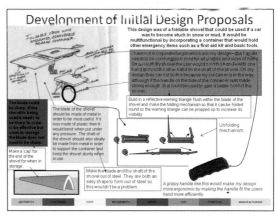

Figure 8.37 Gaining those valuable marks

Figure 8.37 is one of a number of pages taken from a design portfolio, from the development stage. It is an excellent example of work and gains very good marks in all five strands. There is much detail for Strands 2 and 3 and the communication is of a very high standard, with the 3D drawings that are rendered. The work also uses English language well in the annotations.

There is some excellent use of CAD used during the design stage. This is one of several development sheets (in addition to the initial ideas), so this designer will gain top marks in Strands 3, 4 and 5 and very high marks for Strands 1 and 2.

KEY POINT

- Most of the available marks are for your graphical and modelling skills, but there are also marks awarded in Strand 2 for the correct use of English language, spelling, punctuation and grammar.

Figure 8.5 showed research into solving the problem of a bride remembering details about her many guests on her wedding day. Some ideas and developments of solutions to this problem are shown in Figure 8.38.

The range of ideas, some of which are innovative and show flair, will attract really good marks in Strand 1. In this example, there is also a good amount of information addressing Strands 2 and 3. There is a very artistic presentation of ideas, gaining good marks in Strand 4 as a range of skills is used. The ideas are very well presented, colour is used well and 3D presentation is undertaken to very good effect. There is very limited reward in Strand 5, the use of ICT, and the three slides just gain one of the two marks available for the use of word processing. (What you cannot see is that the the designer actually used other ICT elsewhere in their folio in IAO3).

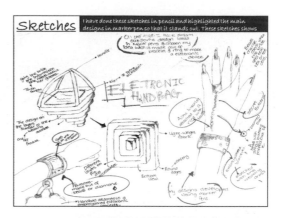

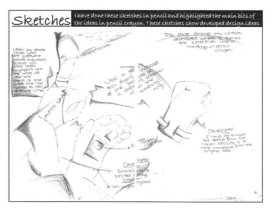

Figure 8.38 Development: the electronic handbag

Figure 8.39 shows some ideas for an MP3 storage device. Marks are gained in Strand 1 and the communication aspect (Strand 4) gains healthy marks. There are some creditable aspects to the comments for Strands 2 and 3, but they are limited. There is limited reward for Strand 5 for the word processing.

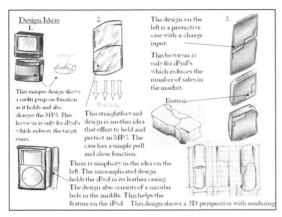

Figure 8.39 **Simple but effective communication**

 QUESTION 4

What additional communication techniques could be used to gain further marks in Strands 4 and 5 for these ideas?

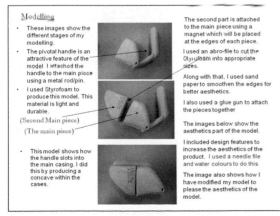

Figure 8.40 **Additional marks**

Figure 8.40 shows some Styrofoam modelling undertaken during the design for the MP3 storage device. It is used to show the folding mechanism and to test ergonomics by actually holding and folding the model. Then the design will be modified.

This is exactly why modelling is used and gains good credit in Strand 4. The comments about the tools used and how the model was made (in red text) will gain no credit.

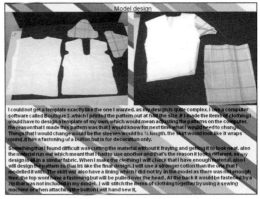

Figure 8.41 **Modelling using textiles**

Figure 8.41 shows a part-made mock-up of a design that has had the pattern produced using ICT. Marks could be gained for the modelling in Strand 4 and also for the ICT in Strand 5, but only if evidence is provided. On this slide, the designer simply explains what she has done, which is not real evidence.

 EXAMINER'S TIP

Screen dumps with annotation are a quick and effective way of evidencing your use of CAD. In IAO3 there are no marks for the CAM aspect of this, but as long as the CAD is evidenced then marks will be gained.

Figure 8.42 shows design work for an item of child's clothing for a special occasion. High marks are gained in Strand 1 for the range and development of ideas, and also in Strand 4 for a range of communication techniques that includes some modelling. There are

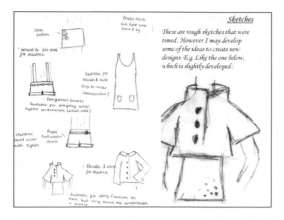

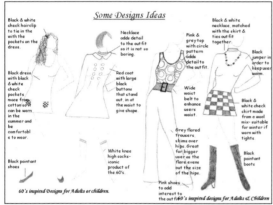

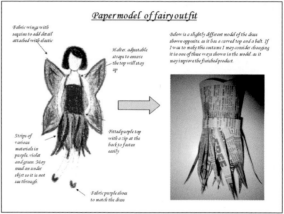

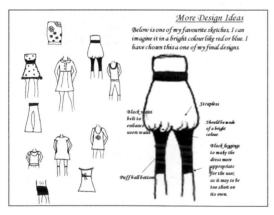

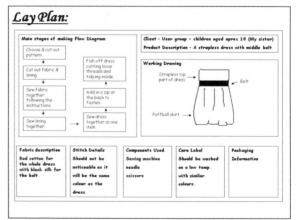

Figure 8.42 A variety of skills

some marks for Strands 2 and 3 but the great strength shown by this set of design folio sheets is in Strands 1 and 4.

QUESTION 5

What marks do you think you will get in Strand 5? Probably more than you might think.

Finishing your work for Unit A551

There is no set way of concluding your work for this unit.

- You may reach a final solution or decision and present it in graphical format. Examples of this are shown in Figure 8.43 – this is really what you should aim for.

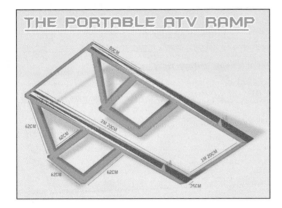

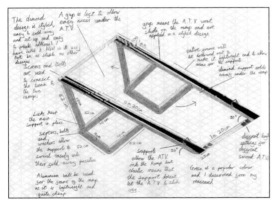

Figure 8.43 **Final solution**

- You might also compare your final conclusions with your original specification and produce a new specification that is more in tune with the user's needs, now you have explored the problem and possible solutions in greater detail.
- You may produce a manufacturing plan to round off your design work.

- It could be that you have moved forward and solved most of the problem you set out to solve, but need to resolve other issues in the light of your design work.
- You might embed some aspects of marketing your designs into your portfolio.

All of these would round off your design activity and could well plug gaps and gain you marks, providing they are giving meaningful information to the examiner.

Reminders and tips

> ### Situation
>
> Camping, travelling, and picnics, places where you might be feeling a little peckish. What better way to deal with it than with some toast.
>
> Currently the only way to make toast is either a toaster that runs on mains electricity or a rack for over a campfire.
>
> However this is not practical on a car journey, when camping without a fire, or in another situation where using mains electricity is impossible.
>
> Instead, a lightweight, small, hygienic, and easy to use solution is required
>
> **Intended users.**
>
> The intended market is anyone who likes toast, i.e. everyone over the age of 2. However due to the practicalities of operating a toaster it will only really be usable by adults and teenagers.
>
> Technology Design Project

Figure 8.44 **Good situation, poor user**

Remember that your starting point is fundamental to your success.

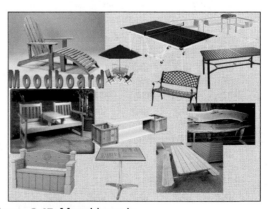

Figure 8.45 **Mood boards**

There are no marks for mood boards but . . .

Figure 8.46 Annotated mood board ... with appropriate annotation they can gain good research marks

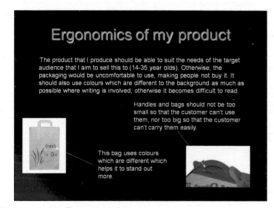

Figure 8.47 Ergonomics

Don't copy irrelevant data into your folio and don't waste time explaining what such things as ergonomics, anthropometrics, consumer law, copyright or patents are. The examiners know what they are; what they need to know is how they affect your design ideas!

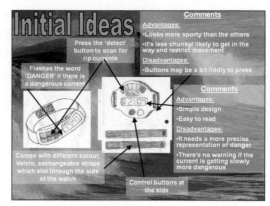

Figure 8.48 Ideas

Produce as many different ideas as you possibly can. The examiner is looking for a range of good ideas.

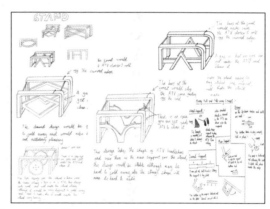

Figure 8.49 Development

Squeeze as much real information in as you can but don't waste time repeating what you have already said. Repeated information cannot gain credit.

Specification point	Idea 1	Idea 2	Idea 3	Idea 4	Idea 5	Idea 6
1	🙂	🙂	🙂	🙂	🙂	🙂
2	🙂	😐	😐	🙂	☹	🙂
3	🙂	😐	😐	😐	☹	😐
4	🙂	😐	😐	😐	☹	🙂
5	🙂	😐	🙂	😐	☹	🙂
6	🙂	😐	😐	🙂	🙂	🙂
7	🙂	😐	😐	🙂	☹	🙂
8	🙂	☹	🙂	☹	🙂	☹

Figure 8.50 Comparing against your specification

Whichever way you choose to indicate how you are using your specification in your designing, avoid things like Figure 8.50. The information given means almost nothing and will gain minimal credit, if any. It is your comments that are valuable.

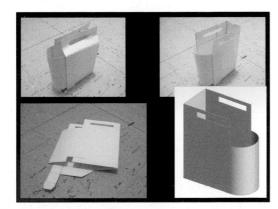

Figure 8.51 Using CAD

CAD can be used for modelling but you can't gain marks for it twice (i.e. in Strand 4 – modelling – and Strand 5 ICT).

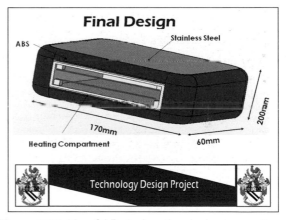

Final Design

Stainless Steel

ABS

170mm

200mm

60mm

Heating Compartment

Technology Design Project

Figure 8.52 Use CAD during the designing and not just at the end

Final free gift

PlayStation 2

Here are all the different free gift I have designed to give away as a free gift with the magazine. As you see some of the free gift are fairly cheap because they are very simple and have no different color. For e.g. key rings clock CD and DVD racks etc... however others I have made are expensive to made and take time like the car remote, control fan ,games.

Figure 8.53 Remain focused

If you design something that does not address your brief or specification, then credit cannot be given.

Remember: You do not have to make what you design in this unit!

CHAPTER 9

UNIT A552: DESIGNING AND MAKING – THE INNOVATION CHALLENGE

LEARNING OUTCOMES

By the end of this section, you should have developed a knowledge and understanding of:

- the structure and purpose of the innovation challenge exam
- how to respond to the different parts of the innovation challenge.

The innovation challenge is an unusual exam. It is designed to test your creative design ability. You will be given a design problem, which is set by OCR. Working through the design process you will develop and prototype a solution to the design problem. The exam will take place in a Design and Technology room, not in the exam hall. Your teacher will give you clear instructions about each part of the exam and about what you are required to do.

About the Innovation Challenge

The challenge has run successfully for a number of years. Students who have undertaken the challenge exam have enjoyed the experience. Yes – they enjoyed an exam! Here are some of their comments:

'More enjoyable than normal exams.'
'A nice calm exam that is not as stressful as other exams.'

Figure 9.1 Students undertaking the challenge exam

'Can we stay in and carry on?'
'I'm really looking forward to this exam.'

The innovation challenge is in three parts. It consists of a six-hour design activity (in two separate three-hour sessions) plus a 30-minute reflection time. In Session 1 you will be planning and designing, and in Session 2 you will be making the model or prototype of your design.

▶ Session 1

This is the first three-hour session of the innovation challenge. Your teacher will explain how the innovation challenge works.

Although the challenge takes place outside the exam hall, you must remember that the rules for external exams still apply. You will be expected to work in silence. At one point in the challenge you will need to present your ideas to other students and you will obviously need to talk at this point! Your teacher will explain to you how this part of the session will be organised. Remember, unless your teacher gives you specific instructions to the contrary, you will work in exam conditions.

The role of your teacher and other staff

Your teacher has two main roles to perform during the innovation challenge. These are:

- to monitor health and safety and intervene if they see any unsafe activity
- to organise the running of the challenge.

In addition to your teacher, you will also have an exam invigilator and possibly a technician in the room. Staff are not allowed to give advice or guidance to you about your design or making activity. They are also not allowed to cut materials to shape for you. Remember,

examiners are assessing *your* ability and not that of the staff around you!

The exam invigilator will make sure the rules for conduct within the innovation challenge are being followed correctly.

Teacher script

Your teacher will read instructions to you from the teacher's script throughout the challenge. The script is used to ensure that the challenge takes place in a structured and well-ordered manner. It will tell your teacher exactly what to say to you at each point in the challenge and will also help them with organising the activity. The script will give specific times for the completion of each section of the innovation challenge. Your teacher may alter these times to fit in with your school or college timetable.

Starting the challenge

Your teacher will read instructions to you about the challenge and how you are expected to behave. They will then read through the innovation challenge task sheet with you and help you to identify which of the four tasks you will undertake.

Each of the four tasks will relate to a single theme. The theme will change annually and will run for both January and June sessions in a calendar year. For example, the theme for 2010 is 'School Sports Days'.

Your teacher will tell you about the theme for the current year.

You will be issued with a workbook that you will use to record all of your design activities. Each section (or box) of the workbook is numbered, e.g. Box 1. Your teacher will refer to these numbers when they are reading from the teacher script.

The following section should be done in Box 1 of the answer booklet			
6 minutes	Box 1. Initial thoughts. Allow 6 minutes with a reminder after 5 minutes.	*Remember, from now on you are creating your own ideas. You should not talk.* *Open your answer booklet and find Box 1.* *The first thing we would like you to do is to put some of your first thoughts down on paper.* *Remember, we want you to be as creative as possible, so sketch and add notes of any ideas you have, even if they seem a bit risky or outrageous at this stage.* *We really want you to feel able to 'let your mind go out to play'.* *In this box put down your initial thoughts. You have 6 minutes, so work quickly. Try and remain focused.*	
The following section should be done in Box 2 of the answer booklet			
8 minutes	Box 2. Allow 8 minutes with a reminder after 6 minutes.	*Look at your initial thoughts.* *Highlight the areas that interest you.* *Think about the challenge you have chosen.* *Which three ideas are worth developing?* *Then:* *Fill in Box 2 with three possible design briefs.*	Explain highlight – draw round, ring, highlight pen, make obvious.

Figure 9.2 Extract from the teacher script

9232562

9232562

OXFORD CAMBRIDGE AND
RSA EXAMINATIONS

General Certificate of Secondary Education
DESIGN & TECHNOLOGY
(PRODUCT DESIGN)

Unit A552 Innovation Challenge

INSTRUCTIONS TO CANDIDATES

You will have a total of 6 hours to complete the examination. This is normally 2 x 3 hour sessions.

At the end of the examination you must have:

- selected one of the challenges detailed on this paper;
- completed an answer booklet showing your creative thinking and how your idea works;
- produced a model/prototype to show the important features of your design;
- have at least four photographs fixed in your workbook showing your modeling activities;
- produced a persuasive argument about why your product will attract the users you are aiming at;
- completed the "Reflection" section of the workbook at some time between 24 and 72 hours after completion of the challenge.

SCHOOL SPORTS DAY

Situation:

On a school sports day, not all students participate in the sporting events.
The opportunities for these students still to be involved and support the event and its arrangement are numerous.

From the lists of challenges below select ONE challenge for which you will design and manufacture a prototype solution.

Challenge One
The school operate a four 'house' system. Students are allocated to a 'house' and represent it when they compete. The crowd are often located long distances from the score board area and are not able to tell which house is winning. A system of easily showing the positions of the four 'houses' throughout the event is required.

Challenge Two
Many students take part in more than one event during the activity. Students can become tired because of this. The school have requested that a healthy hi-energy snack be designed for students to purchase. The snack should reflect the sports day theme.

Challenge Three
The weather at the sports day is going to be hot. All students are going to be given drinks throughout the afternoon. The drinks will be served in disposable cups. A method of carrying large quantities of drinks is required.

Challenge Four
The weather at the sports day is going to be hot. A method of providing temporary shelter for participants in events is required.

Figure 9.3 Sample task sheet linked to the theme 'School Sports Days'

You are *not* allowed to move forward to the next section of the workbook at any time. However, you may go back and add more to sections you have completed earlier in the session. When you are undertaking the 'reflection' part of the challenge you are allowed to read your workbook, but you must not add any more work. You are only allowed to complete the reflection task.

 EXAMINER'S TIP

If you have not got enough space in the section of the workbook you are using, you can use the 'additional space' that is provided throughout the workbook. Work in the additional space section should be clearly labelled to show the examiner which sections of the workbook the work relates to. So, if you want extra space when you are working in Box 5, you will label your extra work 'Box 5 continued'.

The handling and inspirational collections

Having identified the task, your teacher will now introduce to you a 'handling collection' and an 'inspirational collection'. These two collections are designed to help and inspire you with your designing.

The handling collection will consist of items that are directly linked to the theme, as shown below.

Figure 9.4 Handling collections

The inspirational collection will contain items that have interesting or unusual methods of operation, use of materials/ingredients or methods of manufacture and/or assembly. These will not be directly linked to the theme but they could help inspire you to be creative

when you are responding to the selected task. The inspirational collection may consist of products, components or images. Your teacher will talk to you about the 'wow' factor of these items.

You will be allowed to physically examine both the handling collection and the inspirational collection and to ask your teacher questions about them. But remember that your teacher can only tell you about the

 EXAMINER'S TIPS

It would be beneficial to have a copy of the workbook to review whilst you read the next sections. Ask your teacher or download a copy from the OCR website.

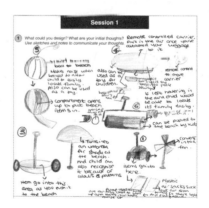

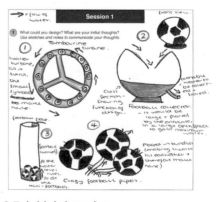

Figure 9.5 Initial thoughts

items in the collection – they cannot guide you in any way. Both collections will be available to you throughout Sessions 1 and 2 of the exam.

Box 1 – Initial thoughts

In this section, you need to record your initial thoughts about the design problem. These are not intended to be fully worked ideas but literally your first thoughts about what you could design to satisfy the task you have been given. You may draw quick sketches or describe your ideas. It might help you to use an analysis diagram to help develop your thoughts.

There is no right or wrong approach to this section. The examiner is simply looking for a range of creative ideas/thoughts that could lead to a solution of the task given. Try to think 'differently' and not suggest obvious things.

EXAMINER'S TIP

Don't be afraid to record your thoughts, however strange or silly you think they may be. They could provide the inspiration for a truly creative solution to the task.

The photos in Figure 9.6 show examples of a creative response to a need for inexpensive children's furniture. Cardboard furniture could easily have been dismissed without any further thought.

Box 2 – Possible design briefs

In this section, you should consider the work you have just carried out in Box 1. Which three ideas offer the most creative and interesting ways of solving the task you have

been set? Which ideas will allow you to demonstrate your design skills to the examiner?

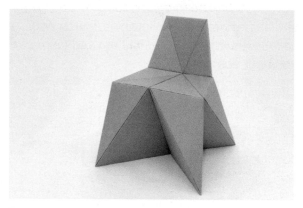

Figure 9.6 Not as strange as it first sounds – cardboard furniture

You will need to record your best three ideas and write a design brief for each of the three design problems you have identified. Refer to the section about design briefs in Chapter 2 of the Product Design Toolbox for help with writing design briefs.

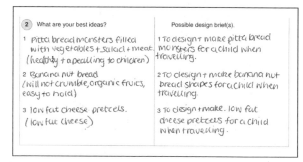

Figure 9.7 Box 2 – possible design briefs

Box 3 – Design brief, user, where will the product be used?

You may not find this page at first! The workbook has been designed to enable you to see this page while you undertake all your future design and evaluation work. The page folds out from the back of the workbook. Remember, you are designing to meet the design brief, the needs of the user and the specification that you record in Boxes 3 and 4. Keeping this page in view will give you a constant reminder of what you are trying to achieve. Keep looking at it to ensure you keep on track!

Figure 9.8 Boxes 3 and 4: design brief, user and specification

Design brief

You will need to select one design brief from the three you have just recorded in Box 2 as your possibilities. Your design brief should be 'open' and allow you the opportunity to respond creatively in your subsequent design work. Don't add too much detail to your brief,

as this will limit your opportunity to come up with a creative design. For example, a brief that says: 'I am going to design and model a children's toy box; it will be a square shape, with four wheels and will be painted blue and green' would really limit your design opportunity.

EXAMINER'S TIP

Think about your possible briefs carefully. Which brief will give you the opportunity to really show off your design skills and creativity? Which brief will you enjoy doing the most?

Clients/users

Who are the intended users of your product? What will they want from the product? How could you make your product appeal to them?

In this section of Box 3 you need to give information about your intended user group. You should use this information to aid your designing. If your solution does not appeal to your target users, your product will not be a success.

EXAMINER'S TIP

The examiner will expect to see reference to your users within your design ideas. They will be able to reward marks if you are seen to consider the needs of the user.

Where will the product be used?

Give a brief description of the environment where the product will be used. Does it create any special design issues such as dust, moisture or noise? How will the product be used?

Box 4 – Specification

The specification is a list of requirements that your design must meet. You will need to think about the work you have completed so far and the task you are undertaking. What are the key features that your design must have? Avoid using general statements such as: 'It must be colourful, it must not have sharp edges, it must be affordable.' This type of point could apply to almost any product. Remember that the word specification means that you need to give specific details about your design.

EXAMINER'S TIP

Do not just reproduce the information that is given in the task within your specification. If you do this, your mark for this section will be low. To find out more about writing a specification, refer to Chapter 2 in the Product Design Toolbox.

Box 5 – Start designing

In this section of the workbook you are required to produce a range of innovative and creative initial design ideas. These ideas should solve the problem that you have identified within your selected design brief in Box 3 and satisfy the design specification you developed in Box 4. Your ideas should not simply be your initial thoughts from Box 1 drawn again, as these do not respond to your chosen brief and specification.

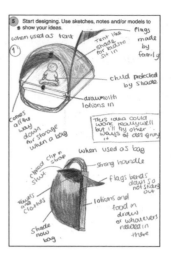

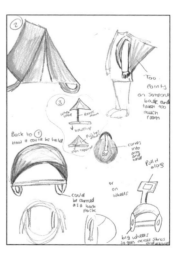

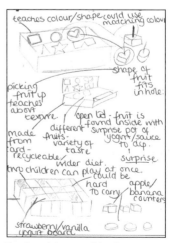

Figure 9.9 Design ideas with clear annotation

You are allowed to use sketches, notes and/ or models to communicate your design ideas. If you use modelling as part of the designing process, it is a good idea to take photographs and stick these into Box 5. Annotate the photographs to ensure that you fully communicate your design proposal. You may need to show more than one view of your design in order for the examiner to understand your idea. Make use of techniques such as rendering, or include material swatches to improve the communication of your design.

EXAMINER'S TIP

Make sure you use annotation rather than labelling to communicate your design thinking to the examiner. Annotation offers an explanation about or justification for a choice of material or feature of the design. A label simply shows the material or feature. The examiner will be assessing your ability to communicate as well as design.

Box 6 – Review of ideas

In this section you are encouraged to review your progress so far and to reflect on your design. You will need to answer four questions:

- What do you think of your ideas so far?
- Which is your best idea?
- Which is your most unusual idea?
- What problems can you see?

These questions will help you prepare for the next stage of the challenge.

Box 7 – Reflect and record and Box 8 – Feedback

This really is a 'different' aspect of this exam. At this point you are required to plan and then give a presentation to two or three other students within your group. You will be able to tell them about your ideas and what you plan to do next. They will be able to ask you questions about your design. They will then make suggestions about how they think your design could be improved. You will need to record these suggestions and decide if you will use them in your future design development work.

This aspect of the challenge exam reflects what happens within 'real-life' situations. Other people will see things in your design that you may not have considered. They may have different skills or knowledge and could trigger a whole new development of your design. It's important that you plan for this activity carefully to ensure that you get the most helpful feedback from others in the group. When it is your turn to give feedback to the other students, you should try to be as helpful as you can.

Box 9 – Developing your idea

You will now have the opportunity to undertake work to develop your design idea further.

The examiner will be looking for evidence of development in the design between your ideas in Box 5 and the final idea that you have at the end of this development session.

You will need to work quickly. You will only have 25 minutes to develop your chosen idea.

EXAMINER'S TIP

You can draw, model or describe in order to develop your design further. If you use modelling, make sure you stick a photo of your model into Box 9 and annotate it to explain your development to the examiner. To award marks in this section the examiner will need to see clear development between your design in Box 5 and your developed design in Box 9.

You will also need to think about:

- size, shape and assembly
- the modelling materials you have available
- the modelling materials that you may need which aren't available
- the components that make up your design
- how the product will function.

If you are working with food ingredients, you will also need to consider taste and nutrition

A photograph (Photo 1) will be taken at the end of 25 minutes to record your progress. Make sure the photo shows all aspects of the development work that you have undertaken.

Box 10 – Questions

This section of the workbook is designed to make you think about your 'developed design' and whether or not you are actually going to satisfy your design brief and specification.

You will need to answer four questions:

- What does your design do?
- What would you like your design to do?
- How could your design become environmentally friendly?
- How does your design appeal to the user group?

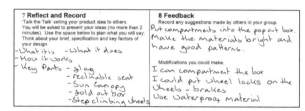

Figure 9.11 Student response to Box 10

Prototyping materials

Your teacher will now introduce the prototyping materials that will be available to you. They may demonstrate some techniques for working with or joining the materials, but they will not be able to help you with your actual modelling. You should think about your design while the teacher is talking. Which materials will be most suitable for you to use? Remember that you don't have to use all the materials that are shown to you.

Figure 9.10 Development of ideas

Figure 9.12 Prototyping materials

 EXAMINER'S TIP

Remember that you are not making a final product. You are modelling a prototype that can be used to test and evaluate your design. The examiner will be looking for an appropriate selection and use of modelling materials.

The next three sections of the workbook (Boxes 11–13) are all related to planning. They will show the examiner how you have considered materials and construction methods prior to starting to make your prototype.

Box 11 – Your model

This section of the workbook is split into two columns. In the first column you need to give the name of the component part for your design, e.g. wheel, leg, waistband, binding agent. In the column next to the name of the component part, you need to give details of the modelling material or ingredient that you will use to make your prototype.

You will then need to identify appropriate methods of joining or combining these materials or ingredients together. Remember that you will only have three making sessions of around 45 minutes each to complete your prototype. Therefore processes that require long set-up times, drying times or cooking times may not be the most appropriate for you.

The final part of this section will ask you to consider how you could use CAM or other electronic applications to manufacture your prototype product.

Figure 9.13 Student response to Box 11

Box 12 – Action plan for Session 2

This is the final section of Session 1. In the next session you will start to manufacture the prototype of your design. You need to think about how you will approach this and plan what you will need to do. Use the space provided in Box 12 to plan your Session 2 activities. Ask yourself the following questions:

KEY POINT

- If you do bring in any modelling materials you *must not* do anything to them until you are actually in the exam. If you modify them in any way before you bring them in they will be taken from you.

- Do staff need to get anything for you?
- Do you need to bring in any materials?
- Do you need to find out anything that will help you next time?

▌ Session 2

In the second three-hour session of the innovation challenge you will be making the model/prototype of your design.

You will have longer periods of uninterrupted time so that you can model your idea. Modelling your design will allow you, your teacher and the examiner to see your ideas in action.

As with Session 1, the challenge takes place in the Design and Technology room, not in the exam hall. You must remember that it is still an exam and the rules for external exams still apply. You should not talk to each other unless directed to do so. As before, your teacher's role is to organise the innovation challenge and to monitor your health and safety. Staff are not allowed to give advice or guidance about your design or making activity. They are also not allowed to cut materials to shape for you.

Your teacher will continue to read you instructions from the teacher script. Remember that you are not allowed to move forward in the workbook before you are told to do so, but can return to any of the sections you have completed earlier and add to these parts.

Box 13 – Further thoughts

You will be given a short amount of time to look through your work from Session 1. Having looked at your work, you will be asked to complete Box 13. This gives you the opportunity to explain any thoughts you have had about your design before you begin the process of manufacturing your prototype. Remind yourself of your design brief and specification. Are you on track to satisfy both of these?

Boxes 14, 15 and 16 – Go make!

You will now have three making sessions of approximately 45 minutes each. You will need to work quickly and efficiently if you are to complete the making of your prototype in the time allowed. At the end of each of the three sessions, a photograph will be taken to record your progress. You should glue these photographs into your workbook as soon as you receive them.

Additional photographs may be taken at any time. You should stick them on to the additional space pages and annotate them. You will also need to complete progress reports within your workbook in Boxes 14 and 15. They will require you to identify any problems you have encountered in the manufacturing process and also to suggest strategies to overcome them. You will then review if your solution to the manufacturing problems was successful. All these activities are timed and your teacher will tell you when to stop and complete each section.

With limited time you will need to plan your work carefully. Try to have alternative tasks available so that you don't waste time waiting for tools or equipment. The making time will go very quickly.

Figure 9.14 Student responses to Boxes 14, 15 and 16

EXAMINER'S TIP

It's important that you complete the manufacture of your prototype. The examiner will be assessing the quality of your final prototype and the range of skills that you have used in the making process. The examiner will also be looking for creative use of materials and also checking to make sure that your final prototype looks like your developed design.

Figure 9.15 Examples of students' prototypes

KEY POINT

- If you don't complete a section you can't get the marks for it!

EXAMINER'S TIP

The examiner is looking for:

- a prototype made to a good standard
- a range of making skills
- creative use of materials
- a prototype that looks like your developed solution.

At the end of the final making session you will be given a short amount of time to tidy your workspace before you complete the final part of Session 2.

Box 17 – Evaluation

When you have finished making your prototype, you need to evaluate the work you have done. It's important that you answer in as much detail as possible.

Think about your prototype. Were you able to complete it? Did you manage to show all the features of your design? Explain clearly to the examiner anything you have not been able to show through the production of your prototype. Explain why this has not been possible. Could you have used any alternative approaches?

You will have four questions to answer in this section:

- What did you want to achieve but couldn't?
- What are the most successful things about your product?
- Why do you think your intended user/s would be interested in your product?
- If you had more time, what would you do to develop your design further?

EXAMINER'S TIPS

This section is about the prototype and your modelling, so think about the design of the product and the materials you used to model your solution.

It's important that you answer in as much detail as possible. Give justification for your answers. Fast Forward allows you to look ahead to the next stage of design development. How would you need to develop your design further in order to make sure that it is successful? You may want to sketch as well as write your response. Remember that you can use the additional space sections if you wish.

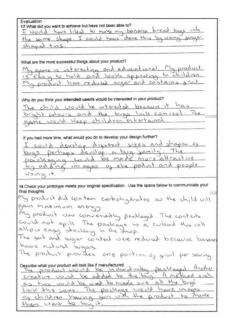

Figure 9.16 Student responses to Boxes 17 and 18

Box 18 – Final thoughts

Before you fill in Box 18 you will need to read your design brief and specification that are in Boxes 3 and 4. Look at your design idea and prototype product. Explain how you have satisfied both your design brief and your specification. If you have not satisfied all of the points, explain why not. Give justification for any answers you give. Fast Forward 2 allows you to explain exactly what your product would look like if it was manufactured and sold. You should consider materials/ingredients, surface finish, colour, texture, smell, function, etc.

EXAMINER'S TIP

Avoid giving short, obvious responses such as 'The design meets all my specification points.' To award high marks, the examiner will be looking for a clear explanation of how you have met the points in your specification.

Time to reflect

Between 24 and 72 hours after completing Session 2 of the innovation challenge you will have the opportunity to fill in page 2 of your workbook, which is called 'Time to reflect'. *At this point, you must not do any further work or make any alteration to any other section of your workbook.*

Time to reflect is not an evaluation of your own performance. It is a chance for you to review your ideas and put forward any further suggestions you may have to improve your product design further.

The examiner will be looking at how you identify any strengths and weaknesses in your design for the product. You should suggest alterations to the design that would overcome these problems. Try to be creative in your proposed modifications. You should use sketching as well as text when working in this section.

EXAMINER'S TIP

A common mistake in this section is to write about the model and the modelling activity, rather than about the product design. Make sure your comments relate to the design of the product and how it could be modified and improved.

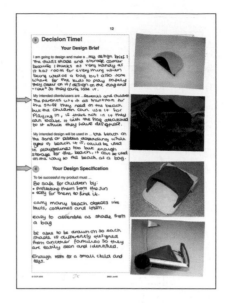

Figure 9.17 Reflection

Figure 9.18 Student response to the innovation challenge

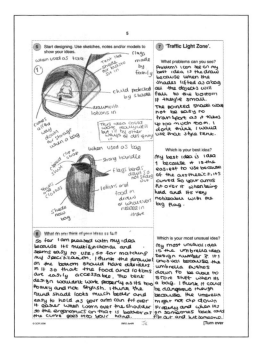
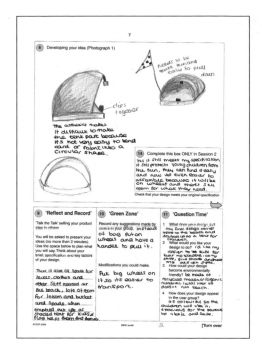
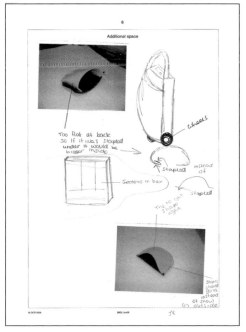
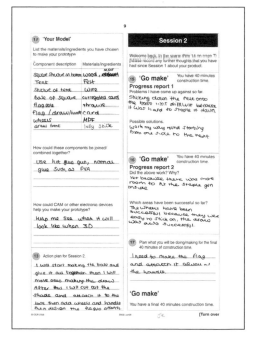

Figure 9.18 continued

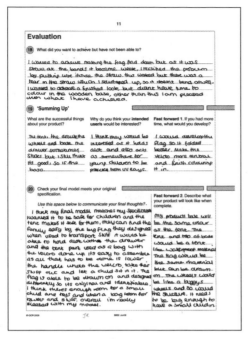

Figure 9.18 continued

ACTIVITY

It is important to develop your ability to work under time pressure. Consider the following design problem:

Transporting beach equipment such as inflatables, buckets, spades, balls, picnic boxes, etc from the car park to the beach can be difficult. A method of carrying beach items is required.

- Write a design specification (time allowed 6 minutes)
- Produce a minimum of three intial design ideas (15 minutes)
- Model one of your designs and develop if further (20 minutes)
- Draw your final developed idea (15 minutes)
- Evaluate your developed idea (15 minutes)

CHAPTER 10

UNIT A553: MAKING, TESTING AND MARKETING

Some people might say that this is the exciting bit. This is your chance to make a quality product. Remember that in this unit you are required to make one complete, fully functional, quality product in any material or combination of materials that is appropriate for the task. A model will achieve no marks in this unit.

10.1 OVERVIEW

By the end of this section you should have developed a knowledge and understanding of:

- The structure of this unit and the evidence required.

Choosing the correct type of project focus and product outcome is essential. You must give careful thought to the type of product you want to make and whether or not you have the skills, materials, tools and equipment to make it. It needs to be interesting to you and others – ideally the product should be as new, as creative and as innovative as possible. The product must allow opportunities for you to use and demonstrate a range of skills that will culminate in a quality product.

No design work is required for this unit of work, but you need something to work from, so although no marks are available, it would be worth making some notes and sketches to

plan how you are going to achieve your high-quality outcome.

A portfolio of evidence is required for this unit. You can present your portfolio using paper or electronic methods but it must not be a combination of both.

Starting points

Your teacher may give you a starting point, and there are lots of other possibilities:

- Your parents or friends could give you ideas.
- You might have seen an idea in a book, magazine or shop you would like to try and replicate.

- You might have seen a design that you think you could improve.

- You may have that 'Eureka' moment – a new and exciting idea.

- You could be influenced by a designer that you are studying, possibly for other Product Design units.

- Your design may be influenced by nature, by shape or by form.

- You could be inspired by something to improve people's lives (such as older or disabled people).

- You may want to develop an educational aid to help the teaching of new ideas.

The opportunities are endless – look around you! Nearly everything around you is a product that has been designed.

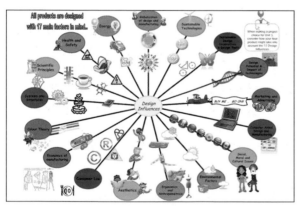

Figure 10.1 All products could be designed with the design influences in mind

Figure 10.2 Examples of products made by students

Figure 10.2 continued

▶ Choosing the product you will make

Think carefully about the product you want to make. Are your manufacturing skills good enough to complete it? Is the product too simple? Will you be able to complete the product in the time available? Does your school or college have the tools, equipment and resources to make the product?

 EXAMINER'S TIP

It would be a good idea to discuss your product idea with your teacher. Are you both happy it is achievable?

10.2 THE PORTFOLIO CONTENT

LEARNING OUTCOMES

By the end of this section you should have developed a knowledge and understanding of:

- the function and structure of the concept page
- the development and presentation of your specification.

For the introduction to your portfolio, you need two things:

- a design concept
- a specification.

Your design concept

This is a way of telling whoever is looking at your product what you are intending to make. This could be in the form of a sketch, working drawing, CAD drawing, words that explain what you are doing, photos or images. What you are making should be clear.

You are not required to generate lots of ideas. You need to show the one, complete, functional, high-quality product that you are hoping to achieve.

EXAMINER'S TIP

Remember the famous saying: 'A picture is worth a thousand words.' A top tip is to start your folio with one picture.

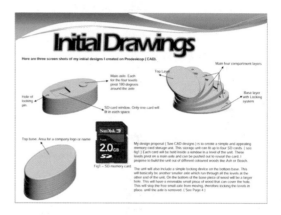

Figure 10.3 Concept page examples

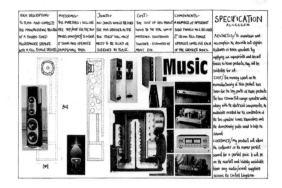

Figure 10.3 continued

▶ Your specification

This is a list of justified points that explain what you want your product to do or to have.

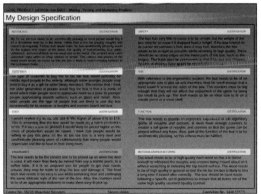

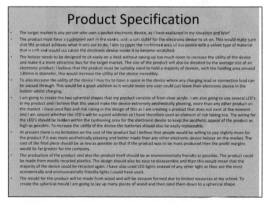

Figure 10.4 Student specification examples

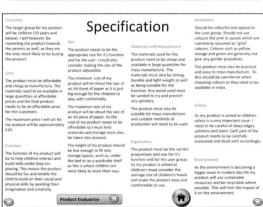

It should include details about the needs of the client(s) and user(s). (Refer to Chapter 2 for more information relating to the specification.)

ACTIVITY 1

Developing your specification – to get started you could mind map all of the possible questions that you might need to ask yourself before you make your product. This will allow you to cover aspects that you might not have otherwise thought about.

Figure 10.5 shows an example of the kinds of questions to ask.

Figure 10.5 Specification questions for a purse or wallet

EXAMINER'S TIP

It is a good idea to make a point, then explain it. This will encourage you to add detail about the specifics of your design and will make evaluation of your product easier at the end.

A good specification helps you produce a product that meets specific criteria usually established through research with and feedback from your user group.

One of the biggest challenges you will have when writing a specification is finding out the needs of your clients or users. The skill of a designer is to create a product that meets the needs of a user at a price that offers good value for money.

In writing your specification, you are setting yourself guidelines to help you check the effectiveness of your finished product. These guidelines will allow clear and accurate evaluation to take place.

Some words do help – A good starting point might be to take the 17 design influences (see Figure 10.1), use them as headings and add notes to help focus your thoughts.

EXAMINER'S TIP

Don't think of your specification as being a rigid set of rules that you can't change. As your product develops, you may want to modify and improve it. This is all part of the ongoing evaluation and should be recorded in your portfolio. If your specification changes, explain why. All good designers are open to change and will adapt their products if a better idea, opportunity or alternative comes along.

ACTIVITY 2

(a) Look around you and choose two products that you enjoy using. Using the design influences as a guide, note down what makes them successful. From your notes, write a specification for the product. How has the designer considered specific design influences and why has this been important for that product?

(b) Now look at your product. Ask yourself the questions listed below. The detailed responses you give should enable you to write a justified specification for your product.

- What is your product intended to do? Simple really, but think about all its aspects, e.g. a chair does not have to be something you simply sit on.
- Are there any safety or hygiene issues related to your product? (British Standards, ways of avoiding injury to the user, etc.)

- What environment will your product be used in? Will the product get wet? Will it be indoors or outside? These factors might have an influence on the design.
- Will your product need to meet any ergonomic or anthropometric requirements? Ergonomics is the study of how a human body interacts with a product, e.g. comfort, ease of use, able to access things easily. Anthropometrics is the average sizes of the human body in a

Figure 10.6 Everyday products for you to consider

given age range. You might look at finger sizes when making buttons for a phone, or at the size of a child's chair in a primary school.

- How can you make the product attractive to your user? Consider the aesthetics of a product. Aspects of aesthetics include colour, shape, size, fashion, texture, smell and sound – all those things that make you like a product.
- Does your product need to be stored? This is very important if you are making food products. Or if you are making a big item, would it benefit from being able to be taken apart or folded to make it smaller?
- How will your product be tested? It is worth considering the best ways to test your product to make sure that it is successful.

Figure 10.6 continued

- Does your product need a protective finish? Some materials will need a finish applied to them to keep them looking good, others will not.
- Are the properties of the ingredients, materials or fabrics used in making the product important? Different materials have different properties, e.g. acrylic is self-coloured and does not need a finish applied. Smart materials can be altered by heat or energy. Your choice of fabrics could be especially important for textile-based products, as they may need to be used outdoors or in extreme heat (to keep someone cool, for example). With an emphasis on healthy eating, choice of ingredients is important.
- Are there any costs you need to consider? Material and selling costs could be important factors.
- Are size and weight a factor with your product?
- How could your product be mass produced? Making a product in quantity is very different from making a product at school. How might this be done in a real-world situation?

Using these questions will give you a good start in forming a detailed specification. There might be other things you can think of that could be added. All products are different and require different criteria depending on their unique requirements.

10.3 IAO4 – PROTOTYPE MANUFACTURE

By the end of this section you should have developed a knowledge and understanding of:

- the production log, its presentation and function
- the impact of sustainability concepts on your product design and manufacture
- hazards and risks, risk assessment and health and safety.

In this section we will look at the requirements of Internal Assessment Objective 4.

Your production log

Production log shows a high degree of skills, use of materials, tools and equipment; images are explained with detail and reasoning; justification of modifications and problem solving during making.	13–18 marks

Extract from the mark scheme

While you are making your product, you are required to keep a production log. This means taking photographs and presenting them in a logical order, or using video clips of all the stages of making, however small, then explaining what is happening in each image. More important, it allows you to explain your

learning more clearly and will be useful when evaluating your product for further modifications at the end of your project. It is really important to do this regularly, so that your production log is as detailed as possible.

EXAMINER'S TIP

Your production log is like keeping a visual and detailed diary.

It is important at this stage that you show what you are capable of. The images need to show the range of skills used, precision and how each stage is achieved. You will need to think carefully about the photographs that you take, as this will be the only evidence of your practical work that will be seen by the examiner. Simply taking multiple photographs that have no specific meaning will offer little or no reward in this section.

You need to explain what materials you are using and why, the tools used in each operation and why they are the most suitable to the job. Explaining what happens in each photograph shows ownership of the work. It demonstrates that the project has been undertaken by you, the student.

If you have never made your product before, the chances are that you are likely to come across problems. It is important within the production log that you show how you solved the problems or modified the design in order for progress to continue. Remember that problems are a good thing – they help you to achieve a good product. Nothing ever runs perfectly. You might even find that you are posed with many questions about your product that you did not even realise existed at the beginning.

The production log is possibly one of the most important aspects of this unit, as it explains to others how you went about creating your product in a safe and economic way. If carried out in substantial detail, this particular piece of work has the potential to allow you to access the higher marks for this IAO. If you are able to include all of this information in sufficient detail, this could get you a third of the available marks for this IAO.

Within the log, show the following:

- how you chose the materials, ingredients or components that are appropriate to the task
- why you chose to use the materials, ingredients, components or fabrics in your product. Your explanations do not have to be lengthy but should demonstrate that you have thought about your options.

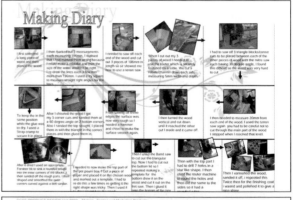

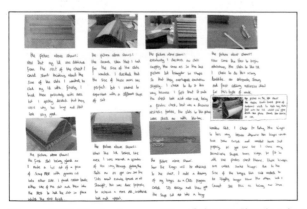

Figure 10.7 Student examples of the production log

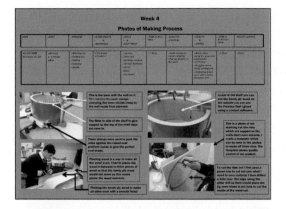

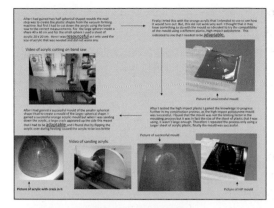

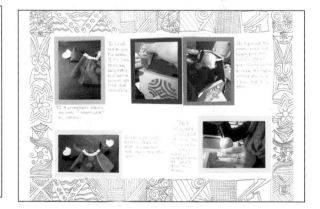

Figure 10.7 continued

Economic use of materials

Shows careful selection and economic use of materials and consideration of sustainability; high understanding of safe working practices; clear understanding of how to achieve precision.	10–12 marks

Extract from the mark scheme

Figure 10.8 Economic use of materials

Why is it important to be economic with the materials used if you are making and selling products? Slight adjustments to a design could save materials and so save money. Considering how components are positioned in a piece of material before cutting will reduce waste. Reducing material usage and reducing waste are good for the environment. The use of sustainable materials could also be considered.

What is sustainability?

Currently, we are using up the earth's natural resources, such as oil or timber, faster than these can be replaced. Global warming is thought to be a by-product of manufacturing processes, energy production and vehicle and plane emissions polluting the atmosphere with chemicals. These chemicals are thought to have caused an increase in the earth's temperature, causing unpredictable weather patterns and melting of the earth's icecaps.

Becoming sustainable involves much more than the current move towards the recycling of household materials (e.g. recycling glass containers). It means a change of attitude for all of us towards the products we buy. We need to:

- replace the natural resources that we extract, leaving the world in a better condition
- take only the resources that we really need
- try to look after the natural world.

We have a responsibility to purchase sustainable products or products that use materials from sustainable resources. Designers and manufacturers have a key role to ensure that sustainable materials are used in there products. They should:

- limit the materials and components used
- use renewable or recyclable materials
- make sure that products are easy to repair, upgrade or recycle
- avoid manufacturing processes that produce toxic substances
- use minimal energy in the production of their products
- consider the effect of distances to transport goods and raw materials.

When designing products it is important to remember the six 'Rs': Rethink, Repair, Refuse, Reduce, Recycle and Reuse. The six 'Rs' provide guidance on how to minimise the damage done to the environment by a product.

What are the six 'Rs'?

Reduce

Designers and manufacturers need to aim to use the least amounts possible of materials and energy in making a product.

Recycle

Designers and manufacturers need to use recycled materials and/or materials that can be recycled after use. Recycled materials are those that can be used again in new products.

Reuse

Designers and manufacturers need to use materials and components (e.g. electronics, cans, bottles, nuts, bolts and screws) that have already been used, and use them again in different products. Designing with this in mind involves making sure that a product can be easily disassembled after it has been discarded.

Rethink

Designers and manufactures are encouraged now to rethink ideas, improving them so they can be energy-efficient, use recyclable materials and generate less waste. For example, some cars are using different fuels to do the same job, but are more environmentally friendly.

Repair

Instead of products just being thrown away when broken or unwanted, designers and manufacturers are making products easier to repair or refurbish.

Refuse

The easiest one of all: you could refuse to buy certain things. But are you willing to give up your music player, fashion goods, cars, scooters, etc.?

REDUCE	Are there any parts in your product that you really do not need? Do you really need to use the amount of material that you have? How could you simplify your product?
RECYCLE	Is your product easy to take apart? What parts of your product could be used again? Would it take much energy to reprocess your product?
REUSE	Which parts of your product could be used again? Has your product got another use without having to process it again?
RETHINK	How can your product do the job better? Is your product energy-efficient? Has your product been designed for disassembly?
REPAIR	Which parts might need to be replaced in your product? Which parts might fail with use or over time? How easy would it be to replace parts?
REFUSE	Is your product really needed? Have you thought about the people who might be making your product – are they treated fairly (pay, living and working conditions, etc.)?

ACTIVITY 3

Look at an everyday product and identify what elements of it could be reused. Can you apply any of the other 6Rs to this product? You could use this type of activity as another starting point for this unit.

Figure 10.9 Safety symbols

▶ Safe working practices

You need to show that you understand that safety is important while making your product. This does not stop at putting on a pair of goggles! What precautions did you take, and why? How did your actions affect the safety of you and those around you?

Things we take for granted, such as sitting at a computer, have risks involved and it is important that you show you are aware of them when you are designing and making your own product.

EXAMINER'S TIP

A good way of demonstrating your safety awareness is to carry out risk assessment.

Hazards and risks

Accidents in manufacturing can have serious consequences to lots of people. Unsafe conditions or acts are called hazards.

There is an important difference between a risk and a hazard. A risk tells you the likelihood that a problem will happen. A hazard is the problem itself. If there is a risk involved, it is likely that it will result in a hazard.

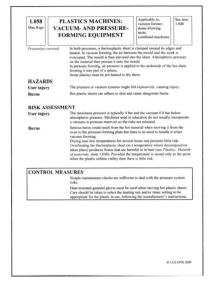

Figure 10.10 Example of a risk assessment form from CLEAPSS®

Assessing the risk

Risk assessment helps to make manufacturing safe by reducing the risk of hazards. Risk assessment is about identifying what hazards there are, how likely they are to happen, and what the result would be if they did happen. Assessing the risks allows appropriate planning to make things as safe as possible. Different processes and areas of the workplace may be identified as being high, medium or low risk.

For example, a risk assessment for the process of sawing a piece of wood might look like this:

What are the risks?

1. You might cut your hands with the saw.

2. Dust from the wood could get in your eyes and throat.

3. You might be distracted by someone else, which could cause an accident.

How can these risks be reduced?

1. The wood should be held securely, allowing hands to be kept away from the work.

2. You must wear a mask and goggles.

3. Ensure there is enough space around you. Display clear warning signs that prohibit others from entering that area when you are cutting.

You should demonstrate your assessment of the risks involved in the practical activities that you need to undertake and suggest how they could be controlled while you are making your product. You could show this in your production log.

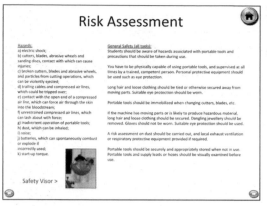

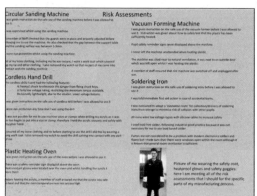

Figure 10.11 Student examples of risk assessment

ACTIVITY 4

Using Figure 10.10 as a template, practise creating a risk assessment for a day-to-day activity. Produce a risk assessment for one of the following: attending a swimming lesson; boiling an egg; crossing a road; using an internet chat room. What risks are involved? Present your findings in a similar table.

KEY POINT

- Remember risk assessment is about identifying what hazards there are, how likely they are to happen and what the result would be if they did happen. Assessing the risks allows appropriate planning to make things as safe as possible; this is something you do on a daily basis without even noticing it!

Precision

To achieve a quality product, it must be well made. A well-made product generally uses good quality materials and shows good workmanship. Good workmanship is all about precision – the small details. So when making your product, did you achieve precision? Often this comes with practice, and/or by using calibrated tools and equipment, patterns, stencils, CAD/CAM, or the application of a finish or presentation.

Working to tolerance

Have you ever tried making something exactly to size? It is unlikely that you succeeded. What you produced would almost certainly have been out by fractions of a millimetre. For you, this may not have been a problem, but in highly complex products a high degree of accuracy is essential to make sure that important parts fit together exactly. The key question becomes: 'Exactly how accurate does it need to be?'

Tolerance is the amount of allowable variation in size from the original manufacturing specification. In some products, such as a jet engine, the tolerance level for each component needs to be very small. The more accurately a product is made, the better the quality in terms of performance and reliability. In products such as a toy plane, the tolerance can be greater. This allows for a wider range of acceptable sizes.

Think about the product you are making. Where on your product are you going to have to address the issue of accuracy and tolerance? At which stages of the manufacturing process will this be particularly important? It might help to make a list:

- Measuring material/fabric or weighing ingredients – incorrect measurement could mean that the artefact or garment is too big or too small, or that the food product is too sweet or too salty.

- Accurate use of CAD for developing templates, jigs and stencils or for developing circuit diagrams – inaccurate drawings and diagrams could prove costly in terms of time, materials and achieving a quality outcome. Components may not fit!

- Marking out, cutting and shaping – inaccuracy here may result in a product that does not fit together or does not store the item that it was intended to store.

Record any evidence of this in your work. This could form a part of the production log.

To achieve the marks for precision you need to include good-quality images in your production log that show your skills off to their best. Close-up photographs are a good technique to use.

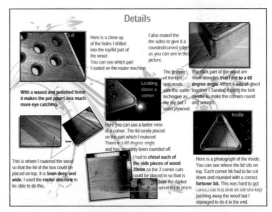

Figure 10.12 Student examples that highlight quality and precision

Presenting your final product

The product will be completed to a high standard and will fully meet the requirements of the design specification.	18–25 marks

Extract from the mark scheme

You have worked hard and completed your quality product. It looks great (even if you say so yourself), but how do you get this across to others?

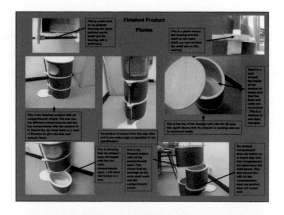

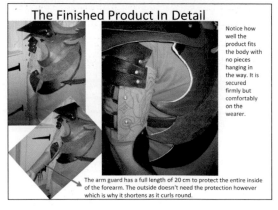

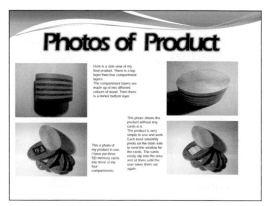

Figure 10.13 Student examples of finished products

This unit of work is known as a postal moderated unit – it means that only your folio is sent away to the examiner. Your practical work will only be seen in the form of a series of photographs/video. These images should show the product from a range of views, the product's features and the quality of the final item.

EXAMINER'S TIP

If you have no image of your final product you will be awarded zero marks. It would be a shame if all that effort and work went to waste! Include familiar items in the images (e.g. a ruler, watch or coin) so that people can gain an idea of scale.

10.4 IAO5 – TESTING, EVALUATING AND MARKETING

LEARNING OUTCOME

By the end of this section you should have developed a knowledge and understanding of:
- product testing, evaluation and improvement.

What's important about this stage of the project is that it's all about taking your product forward. The practical aspect is now complete. No further reference needs to be made in relation to making the product.

So, you have got this great product. How do you find out if it's any good?

Test and evaluate

Evidence of thorough testing by a user/user group and full evaluation with reference to the design specification using both graphical and written techniques of a high standard and structure, with few errors in spelling, punctuation and grammar.	6–7 marks

Extract from the mark scheme

At the beginning of this unit, you wrote a specification. You now need to evaluate how you got on. Explain how you achieved each specification point, and if things changed, why this happened. Remember that nothing is perfect and that it's a good thing to be able to recognise where things can be improved.

Make your points meaningful and be honest about the changes that you have made and that you would make. This will help when you come to improve and modify your design.

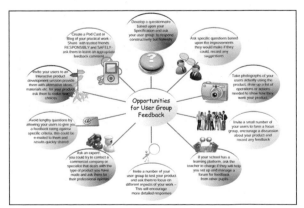

Figure 10.14 Opportunities for user group feedback

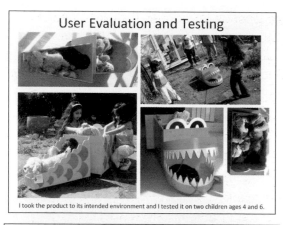

User Evaluation and Testing

I took the product to its intended environment and I tested it on two children ages 4 and 6.

EVALUATION AGAINST SPECIFICATION

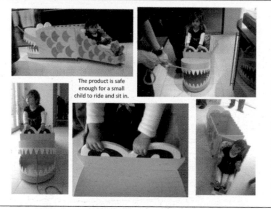

The product is safe enough for a small child to ride and sit in.

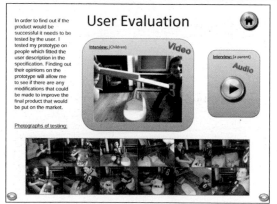

User Evaluation

In order to find out if the product would be successful it needs to be tested by the user. I tested my prototype on people which fitted the user description in the specification. Finding out their opinions on the prototype will allow me to see if there are any modifications that could be made to improve the final product that would be put on the market.

Photographs of testing:

USER EVALUATION

SPEAKER QUESTIONNAIRE

Figure 10.15 Examples of students' testing and evaluation

 EXAMINER'S TIP

Use a spell checker or get someone to look at your work, as accuracy is important here for spelling, punctuation and grammar.

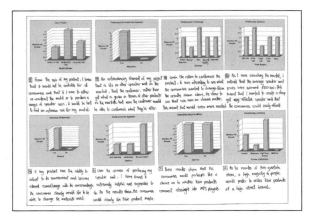

Figure 10.15 continued

EXAMINER'S TIPS

Test your product with your user group. Ask people their opinion. Get your chosen user group to try your product. Does the product work? Do they like it? In your portfolio show how your product was tested using photographs or video. try to summarise peoples' feedback and present it in your folio.

▶ Modifications and improvements

Design modifications/ improvements of the final product are suggested in full detail.	5 marks

Extract from the mark scheme

Look around you: see how products have developed over time. Any product, however good, can be improved or modified, though modifications are not always for the better.

Take a product we are all familiar with – a phone. There is a large market for

accessories for phones (lanyards, phone charms, cases, socks, etc.) All these additions have been thought of as ways to modify or personalise a product and therefore enhance profit opportunities, but not all the additions and modifications have been a success.

When suggesting improvements to your product, give detailed and justified comments, particularly when you consider your user group or target market.

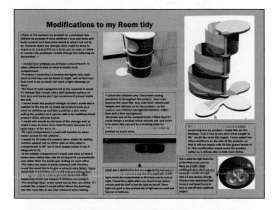

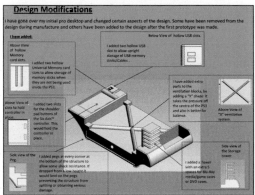

Figure 10.16 Modification – examples of students' work

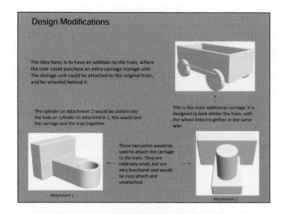

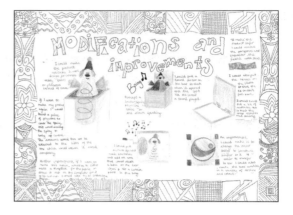

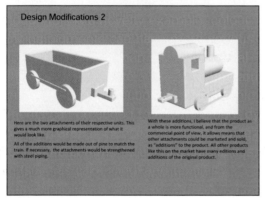

Figure 10.16 continued

ACTIVITY 5

(a) Choose a product (e.g. a calculator, pizza, birthday card, trainers, socks, duvet, steering wheel). List a number of ways to improve the product and sketch a possible modified outcome. Annotate your idea.

(b) Think of ways your product could be improved, enhanced or modified. Try to be innovative in the way you do this. Consider sustainability – could your product be modified or improved using the six 'Rs'? You could refer back to the design influences or your specification. If you could make your product again, what would you do differently to make it even better?

Treat this as a design type of activity. Use notes and sketches, photographs or CAD modelling to show how the product could be modified or improved.

EXAMINER'S TIPS

If you have completed some effective testing with a user group, their feedback should give you ideas about improving your product. Did any issues arise when you were evaluating against your specification?

Remember that you will need to evidence the feedback that the users have given you in your work – make this meaningful by allowing them to use your product for its intended function.

10.5 REAL-WORLD MANUFACTURING

By the end of this section you should have developed a knowledge and understanding of:

- commercial production and quality control methods.

Consideration of quantity production leading to a detailed description of a suitable quantity manufacturing system including details of chosen materials for the main component(s).	5 marks

Extract from the mark scheme

Batch production

Batch production involves the production of batches of similar products. The term 'batch' refers to the scale of production, from a few items to several thousand. These systems are usually flexible and allow a range of products to be produced.

Continuous flow

Continuous-flow production systems take in the raw materials at one end of the factory, and the finished product comes out the other end. These kinds of factories are usually fully automated and run '24/7'. Steel and paper are produced in this way.

Repetitive flow

Repetitive manufacturing is used to produce products that are unchanged over a long period of time. A specific quantity is produced at a specific rate over a defined period of time. This production will be used to produce food items like chocolates and drinks.

Zero defects

The need for consistent quality in products is very important. A company needs to make sure it does not develop a reputation for manufacturing products that are unreliable. It will also want to minimise the costs of rejects and repairs. Gauges, jigs, templates and measuring systems (e.g. callipers, micrometers, electronic sensors) are used to check accuracy.

As a consumer, would you buy a product that

was less than acceptable? You would expect to be sold a quality product that is just as good as the rest.

Using CAM systems greatly helps to achieve fewer defective products but only if the original information that is programmed in through the use of CAD is accurate.

Quality control

Quality control involves different ways to check products as they are made. In most cases, it would take too long to inspect every item on a production line. For this reason, a sample is checked – one in every 1,000 perhaps, or once every five minutes, depending on the product.

The results of the tests are recorded and compared. If defects occur too often, the production process is stopped. This feedback helps to identify problems before too many substandard products are produced.

Some quality control checks will be undertaken and recorded by production workers. Computer-controlled systems are frequently used for machine processes. Computers can supply feedback quickly.

Quality assurance

Quality assurance is the overall approach that a company takes to keep standards high in all aspects relating to a product. Quality assurance provides a guarantee to consumers that the product is a certain standard and will consistently perform to that standard, even after multiple uses.

You would be surprised what some of these tests might look like. For example, vacuum cleaners are literally thrown down flights of stairs a number of times to test their capability of standing up to everyday life.

Ask yourself these questions about your product:

- How accurately made do the different parts need to be?

- Which components need to fit together most accurately?

- At what stages of manufacture would you recommend that a sample of your product should be measured for accuracy?

- What inspection and measurement tests could be carried out?

- How often should the tests be done?

ACTIVITY 6

Think about the product you have made: How will you demonstrate quality control in the manufacturing process? Record your thoughts in your portfolio.

KEY POINT

- Quality control systems help manufacturers reduce wastage and delay in production. They do this by predicting failure before it happens.

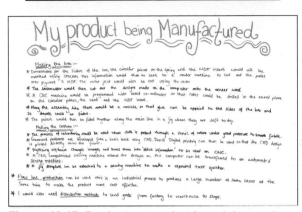

You have just completed your project in school. You need to now explain how your product would be manufactured in quantity in the 'real world'. This needs to be specific to your product and is not just general theory.

How would materials be sourced or made? Break down the product into specific, individual parts and explain how these would be sourced/manufactured, before coming together to be assembled.

What techniques are available to industry that are not used commonly in schools, but achieve the same outcome just on a larger scale? You might already have some equipment in schools that is used in industry.

Figure 10.17 Student example of applying real-world manufacturing techniques to their product

ACTIVITY 7

1   As preparation for this section it is important to research how products are made. This may not be as easy as it sounds, but with the use of the internet anything is possible. Research how a magazine or newspaper is manufactured in quantity. How are the paper and inks made? What is the print process? How is the final magazine/newspaper put together? This is a really useful exercise in understanding how things are made.

2   Find out how a product of your choice is made and present it to your class. A good starting point for this is to use a resource like a video site such as 'YouTube' or 'Discovery Channel' that have lots of short videos showing how things are manufactured in quantity in the real world.

10.5 YOUR MARKETING PRESENTATION

By the end of this section you should have developed a knowledge and understanding of:

- how to market a product, considering the ways in which products and services are sold to consumers.

Marketing presentation is thorough, fully explained and uses an innovative and persuasive approach.	13–18 marks

Extract from the mark scheme

The final stage of this unit is to present your product in an interesting and persuasive manner to one of the following individuals or groups:

- prospective manufacturer
- supplier
- company buyer
- retailer
- end user or user group
- consumer.

Think carefully about the audience and purpose of your presentation. What information about your product and its market do you need to communicate? This should be seen as a 'sales pitch': an interesting way to bring the product to the attention of your end user.

Consider programmes like *Dragons' Den* and *The Apprentice*. How are products advertised? How effective this is? What makes products successful? Think about the technology we have in today's society, e.g. PDAs, multi-functional mobile phones, Wi-Fi,

Figure 10.18 shows some adverts that were found by students. They are very innovative and may help you when thinking of ways of marketing your own product. These images are very inspiring and show what can be done with some imagination.

ACTIVITY 8

Present your favourite advert to your peers. Explain why it is your favourite and why it appeals to you. What makes it effective? Is it novel, funny or thought provoking? Advertising is a strong tool and can influence what you do.

Here are some other options to consider:

- Add SMS and MMS advertising – messages sent when customers walk past the shop window.
- Online questionnaires about the product with the possibility of winning prizes, e.g. earn i-points when buying certain products or services.
- Trial products posted through doors or in magazines. User group participation, in marketing and testing panels for products.
- Product placement, in films and in TV shows.

Figure 10.18 Adverts

- Celebrity endorsement of products could include personalities and/or music.
- Cryptic adverts – get people talking about them until the big reveal for the actual focus of the advert.

- Gimmicks – characters that are immortalised in a cuddly toy.
- Spin-offs – that can cover adverts, TV shows, etc.

ACTIVITY 9

Think about everyday products that you are familiar with, for example a deodorant ad campaign, confectionery or a sports brand – what makes these memorable for their consumers? Can you identify the user group that these products are aimed at? How do they appeal to these particular users? Have you bought products as a result of being tempted by the way in which they have been advertised to you?

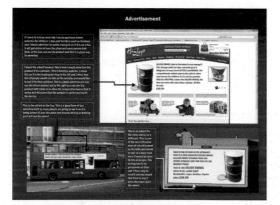

Figure 10.19 Student marketing presentation examples

EXAMINER'S TIP

You will need to think carefully about your target market, user or consumer. Investigate their lifestyle, to make sure you choose a method of advertising and marketing that would be most effective for them to find out about your product.

Portfolio section	Questions to consider	✓
Production log	Do you have lots of images showing how the product was made? Have you taken ownership for the images and explained in detail what you have been doing? Are all tools, equipment and processes explained? Have you shown how you might have solved problems while making Can you show how you achieved precision in your work? How can you show you have been economical in use of materials? Do you have images that highlight the quality of your product? Do the images show all the functions of your product? Are your images of a high quality, in focus, clear, with a variety of angles and show scale?	
Evaluation and Testing	Do you have evidence that your product has been tested in context? Do you have evidence your product has been tested by your user group? Have you evaluated all your specification points? Have you checked your spelling, punctuation and grammar in this section?	
Modifications and improvements	Has your testing and evaluations highlighted the need for improvements or modifications? Have you shown with sketches and notes how your product could be improved?	
Real World Manufacturing	Can you show how identified components of your product can be manufactured in the 'real world'? How might they be made in a factory?	
Marketing Presentation	Is your presentation aimed at a particular audience? Why have you chosen this style of presentation? Is your presentation innovative and different? Does your pitch sell your product? Does your presentation grab your attention? Was it fun to do?	

Table 10.1

This table is a checklist to help you check that you have covered all the areas necessary to complete a successful portfolio.

CHAPTER 11

UNIT A554: DESIGNING INFLUENCES

11.1 THE WRITTEN EXAM

By the end of this section you will:

- be able to answer exam questions related to designing influences
- be able to answer exam questions on iconic products and trendsetters
- be able to design 'in the style' of a trendsetter
- be able to use the knowledge and understanding you have gained from product analysis and other research activities

About this exam

Before we go any further there are some important facts you need to know about this particular exam.

First, your paper will be 'scanned' and then marked 'online' by a specialist examiner. There are two things which you **MUST** remember about this type of exam paper marking:

1 There is a series of 'L' brackets in the corners of all the pages in the examination paper. They are like a 'border' on each page and you must not write or draw outside of the 'border'. If you do, whatever you write or draw will not be seen by the examiner.

2 At the bottom of every page there is a bar

code printed that is unique to you and that examination page. You must not write on or near the bar code or the scanning process will be affected. This will then mean your paper will not be able to be marked in the normal way and will have to be posted to a different examiner. This all takes time. You might not get your results on time because of this.

You will have one and a half hours to answer all the questions in the exam paper. There are two sections to the paper: Section A and Section B. You should spend approximately 45 minutes on each of these two sections.

Section A

There are three questions in Section A and each question is worth 10 marks. The parts of each question have the mark allocation shown in squared brackets [1], [3], [6] etc. These marks are a very important guide and will help you to structure your answers. If a question has 3 marks [3] you must be prepared to give the examiner three pieces of information. It might be that you state a particular point and then explain something about it by giving two additional pieces of information. Equally, it might be that you give two points of information and then some further information. It could also be two examples and some further details. Just remember that if there are three marks you will never gain the full three marks unless you 'earn' them.

Section A questions will have a drawing of a product or possibly two similar products and you will be asked questions about the product(s). You might not have actually done a product analysis activity on that particular product but you will know what the products are, and if you have done other product analysis activities you will be able to start to answer the question.

Figure 11.1 Avoid answering solely from your observations of the drawing

 EXAMINER'S TIPS

Be aware of how limiting one word answers can be. Even if a question has just one mark [1] often one word does not say enough to gain that mark.

Avoid using any of the following terms unless you can really clarify them:

- **light/very light/lighter**. Does this mean light in weight or light in colour? You will have to give some further explanation or details so the examiner knows exactly what you mean.
- **cheap/very cheap/cheaper**. You cannot possibly justify if something is 'cheap' or not. Just avoid using the term.
- **strong/very strong/stronger**. You will have to qualify anything to do with strength and that is very difficult to achieve.
- **tough /very tough/tougher**. As with strong it so difficult say anything meaningful without some real facts to explain what you mean.

Also avoid using generic terms such as metal, plastic, wood, cloth, etc. You will need to be specific such as stainless steel, polythene, plywood or denim to gain marks.

What you must avoid doing is looking at the drawing and answering solely from your observations of the drawing. The images are there just to focus you and make sure you are thinking about the right sort of product. They are general questions about aspects of the product or implications of the product for designers, manufacturers and consumers.

The first part of each question will be the easiest and the different parts will get more difficult as you work through the question. So don't panic; if you find the last part hard, it is meant to be that way.

Question 3 is a little different and will ask you to compare an older product with a more modern one. You will be asked to suggest **three** different successful features of the modern product compared to the older one. And remember that the drawings are there just to point you in the right direction; it is not a 'spot the difference' between the images!

You will then explain about the successful features you have identified.

Section B

There are two completely different styled questions in Part B: Question 4 is worth 10 marks [10] and Question 5 is worth 20 marks [20].

Each year OCR will publish a list of five different trendsetters and an iconic product associated with each of the trendsetters.

KEY TERM

TRENDSETTER – a designer or other body who was responsible for the creation of something that has had a major impact on society in some way or other.
ICONIC PRODUCTS – products that many people recognise as having had a significant impact on society in some way or another.
ERA – a period of time when something took place
MOVEMENTS – groups of people who had a common view or vision and generally worked together to move their vision forwards.

Your teacher will have details of the current eras and movements, trendsetters and iconic products that you will need to study. They will guide you and suggest which are the best one(s) for you to look at. Remember that they change each year.

You can independently find out what they are each year by going to the OCR website. They will be in the format shown in Table 11.1.

	Eras and movements	Trendsetter	Iconic product
1	1960s	Alex Issigonis	Mini Cooper
2	(1970s) Hippies	Yves St Laurent	Kaftan
3	1950s and 1960s	Andy Warhol	Campbell's soup painting
4	1980s	Infra-red light emission	TV remote control
5	1960s	Vegetarianism	Nut loaf

Table 11.1 Examples of eras and movements, trendsetters and iconic products

KEY POINTS

- You must be very careful that you study the correct trendsetter and iconic product and also that you research the types of things the examiner is looking for and not just the things that interest you. It is not 'down to you' to undertake the research. Your teachers must have significant involvement.

- Your teachers have been asked to help you to look at two of the five trendsetters/iconic products. This is to allow you some flexibility and choice when you answer the two questions in Section B.

Section A – A more detailed look

Look at Figure 11.1 again.

The question asks you to give **three** important design features of denim jeans and provides a space for each of your answers.

Feature 1. .... [1]

Feature 2. .... [1]

Feature 3. .... [1]

EXAMINER'S TIP

There is normally more than enough space to write your answer. If you need 'extra' space you are probably missing the point of the question.

Even if you do not own a pair of denim jeans you will know that they will come in range of different sizes, have belt loops, have pockets, are washable, etc.

This is a pretty easy start to the question which will then go on to ask about other factors associated with denim jeans and then other products in a more general way. These will all be based on the different design influences (see Chapter 7 for more details) that you should have taken into account when you did your product analysis activities and indeed your designing unit if you have already completed A551: Developing and Applying Design Skills.

EXAMINER'S TIPS: THINGS TO AVOID:

- Giving theory definitions such as explaining what ergonomics is or what copyright is. The examiner will be looking for how it actually applies to the products in the question and products in general.

- Making things up! It is very obvious to an examiner if you make something up. You will have been taught to move on to another part of a question if you don't know the answer and come back to it later. That's a good plan and at that stage if you are still not sure about an answer then make an educated guess, but don't just make things up.

- Repeating the question in your answer. It wastes so much of your time and gets no marks.

- Answering what you want to answer rather than answering the question.

You already know about the difference between stating something, giving something and describing and explaining. Just remember the marks and that will keep you on track.

Question 3 is a little different to questions 1 and 2. And there are six pointers to help you to make sure you gain as many marks as you can.

Fruit Smoothie   Fizzy drink

Figure 11.2

Point 1

This question will have two drawings of products: one an older or more traditional one; and a more modern and advanced one.

You must always assume that the modern one has some more successful features than the older one. Don't try to get clever and argue that the older one was better in some way. It might well have been better in some respect but that is not what the question is all about.

Point 2

Don't repeat the successful feature which has already been given to you. Identify three *different* points.

Point 3

Remember what was said about the drawings. The drawings are there just to

point you in the right direction; it is not a 'spot the difference' between the images question. Apply your product analysis skills.

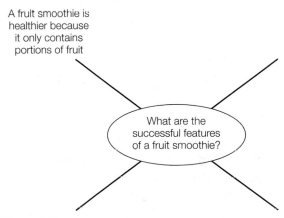

Figure 11.3

Point 4

Look at Figure 11.3. The first point has been given to you: 'A fruit smoothie is healthier because it only contains portions of fruit'. It would still have gained the mark if it had just said 'A fruit smoothie is healthier'. There are no real details or explanation needed in this part of the question, Part (a). Your explanation and details come in Part (b), where you will explain the three *different* points you identify in Part (a). Don't waste time and energy writing things out twice!

Point 5

Part (b) for this example says 'Health advisers recommend that young people, in particular, should eat at least five portions of fruit per day as part of a healthy diet.' It is a two mark [2] question. Can you see the **two bits of information** given? In fact it is a pretty good answer isn't it?

Point 6

Guess what is wrong with this example answer to the same question:

Part (a) *'The fruit smoothie has a screw top'*.

Part (b) *'So the smoothie does not spill if the container tips over and wastes the drink'*.

Yes, you've got it. That answer is about the container and not the smoothie. Sorry.....no marks!

Section B – A more detailed look

So half-way through the exam and we are on to Question 4. If you have listened well in your examination preparation lessons and carried out some additional research on your own then you will have no problems at all with this question.

KEY POINT

- Part (a) of Question 4 will also assess your written communication. Your spelling, punctuation and grammar will be looked at by the examiner.

Question 4 has two parts, (a) and (b), with Part (a) being worth six marks and Part (b) worth four marks.

Part (a) is all about the trendsetter and not the iconic product. It will ask you to choose a trendsetter from the list given and explain why they or it was so important. It is a simple as that. The advice below will help you gain as many marks as you can:

Point 1

Make sure you choose one of the trendsetters from the list on the examination paper and not another one you might have studied or know about. You might know everything about Harry Beck and the London Underground Map but unless he is on the list there will be no marks for writing about him.

Point 2

Make sure for Part (a) you talk about the trendsetter and **not** the iconic product. You will not gain marks if you talk about the iconic product in this part.

Point 3

Actually explain the importance of the trendsetter; do not fill the answer space with irrelevances. Where the trendsetter was born, when they got married and what their children were called probably (almost definitely) has no bearing on why the trendsetter is seen as important.

Point 4

Do not waste time saying things like 'she was a very important woman', or 'he was an important designer'. Get to the point and tell the examiner some 'real facts'.

Point 5

Try to incorporate other products or designs that the trendsetter was involved with if you possibly can. Don't make it up if you don't know, but if you do know then it shows how important the trendsetter was.

Point 6

Don't give your own personal opinions. Stick to the facts and what you have been told and learnt about the trendsetter. Whether you liked them or their work is totally irrelevant and will gain you no marks at all.

Having a structure to your answers

Examiners look for clearly explained, relevant points. The best way to approach this is to begin with a quick mind map. (Use the side of the exam paper.)

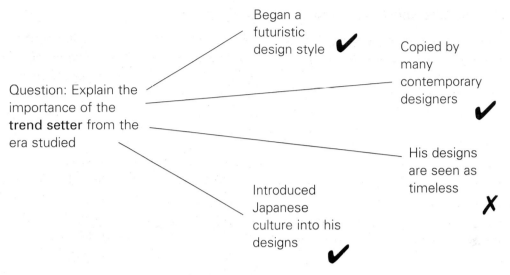

Figure 11.4 Mind map

Having listed as many points as you can think of, you now need to pick the three or possibly four most important points. Once you have done this, it is relatively easy to write a short answer that will gain you full marks.

So for this example an answer might read:

'This designer is important because he began what has become known as the futuristic design style. This design is so influential that even today modern designers copy his style. One of the other influential elements of his work was the way that he introduced Japanese culture into his designs. His work…' and so on.

This approach can also be used for Part (b) of Question 4.

Part (b) is all about the iconic product. It will ask you to choose an iconic product from the list given and explain why they or it was so influential. Again it is a simple as that. Below are six points to help you gain as many marks as you can:

Point 1

Make sure you choose one of the iconic products from the list on the examination paper and not another one you might have studied or know about. You might know everything about Mary Quant and her hot pants designs but if the iconic product is the mini skirt and you talk about hot pants you will get no marks.

Point 2

Make sure for Part (b) you talk about the iconic product and not the trendsetter. You will not gain marks if you talk about the trendsetter in this part.

KEY POINT

- You do not have to link the two parts of this question if you don't want to. So for Part (a) you might choose one trendsetter and for Part (b) a totally different iconic product, which is unrelated to the trendsetter. As long as it is on the list this is OK.

Point 3

Actually explain the influence the iconic product has had and do not fill the answer space with irrelevances. Stating that you like the product or you have been influenced by it has no relevance to the question at all.

Point 4

Do not get side-tracked and start drifting towards talking about the trendsetter. It is very easy to get confused at this point in the paper.

Point 5

Try to relate to other products or designs which have been influenced by the product you have chosen to show how influential the product actually was.

Point 6

Look out for the 'focus points' written for this part of the question. It might ask you talk about the materials, the function, the style or some other factors associated with the iconic product.

You have done your research with your teacher, probably done some on your own and you know about the designer and the products. Now you move on to the last question which has a total of 20 marks [20] – that's one third of the marks for the whole paper!

Question 5 – The design question

To do well in this question you will need to be able to do the following:

- sketch, draw and annotate to score well
- develop design ideas. 'Develop' means improve upon an initial idea
- 'evaluate' design ideas. 'Evaluate' requires you to critically compare something against a list of criteria (e.g. a design specification)
- not waste time re-drawing your ideas as you move through from ideas to development to evaluation. Designing is all about moving forwards and changing things for the better and you need to be able to do this automatically
- not use colour in your answers. The examination papers are scanned in black and white. Using coloured pencils will not show up on your examination paper. So, if you decide that it is important that a component part of your design is coloured 'red' then you must annotate this and explain why it needs to be red. This will also save you a lot of time when you are designing
- write a clear specification. The specification is not only four marks [4] for Part (a) of Question 5 but affects the marks in all other parts of the question too. The specification must give 'SPECIFIC' points for the item you are designing. It will give structure to your designs and will give detailed points against which you can carry out the evaluation.

Figures 11.5–11.7 show you an example of a candidate's responses to a design question.

The trendsetter for this question was 'Isambard Kingdom Brunel' and the iconic product was the SS Great Britain.

DESIGN SITUATION

A child's toy inspired by the work of Isambard Kingdom Brunel.

The first stage is to write four specification points. As with all design specifications, the points must be specific to the product.

EXAMINER'S TIP

General specification points such as 'it must be healthy' or 'it must not be heavy' will score no marks. Instead, a specification point such as 'it must contain at least 5g of fibre' is what is required. A good answer to this question is given below.

Part (b) requires that you sketch a range of ideas. The examiner will expect to see three or more 'different' ideas. So make sure you produce plenty of different ideas and your comments (annotations and not labels) relate to your design specification. A good example is given in Figure 11.6.

Part (c) is about developing one of your ideas. This means that you need to change it or modify it in order to improve the idea. The most important thing here is that every change you make is justified in terms of your four design specification points. A range of sketches with annotation is required.

Use phrases like 'specification point 3: I have added a zip to the front of the hat, so that it can be opened'.

A good example is shown in Figure 11.7.

Identify **four** important specification points for your chosen design situation.

1 The Aesthetics of the toy, making it bright and colour, attracting to a child. [1]

2 The shape of the toy, how it would fit comfortably in a childs hand. [1]

3 ~~The~~ How safe the toy is for a child to use with no sharp edges at all. [1]

4 How durable the product would be and the lifespan it would have. [1]

Figure 11.5 Student response to Question 5(a)

(b) Use sketches and notes to show your initial ideas.

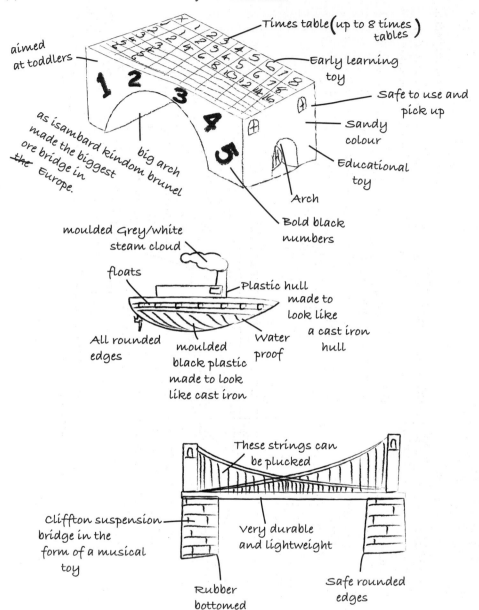

Figure 11.6 Student response to Question 5(b)

(c) Use sketches and notes to develop one of your ideas.

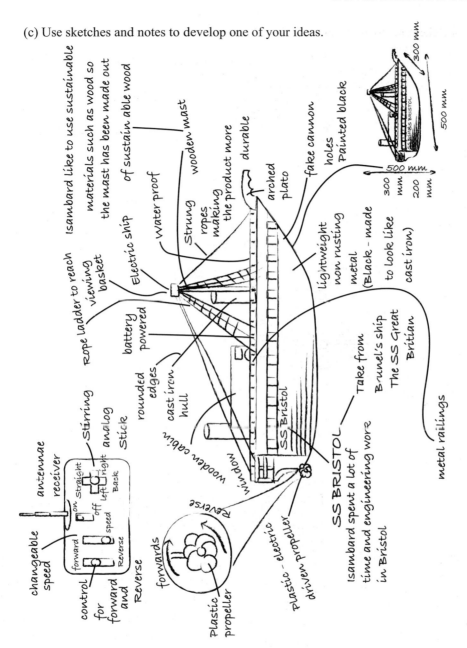

Figure 11.7 Student response to Question 5(c)

For Part (d) you need to evaluate your design against each of your specification points. Take each point in turn and make clear statements that clearly show how your design has met the specification point.

(d) Give details of your final proposal showing how it meets the four specification points you identified in part (a) of this question.

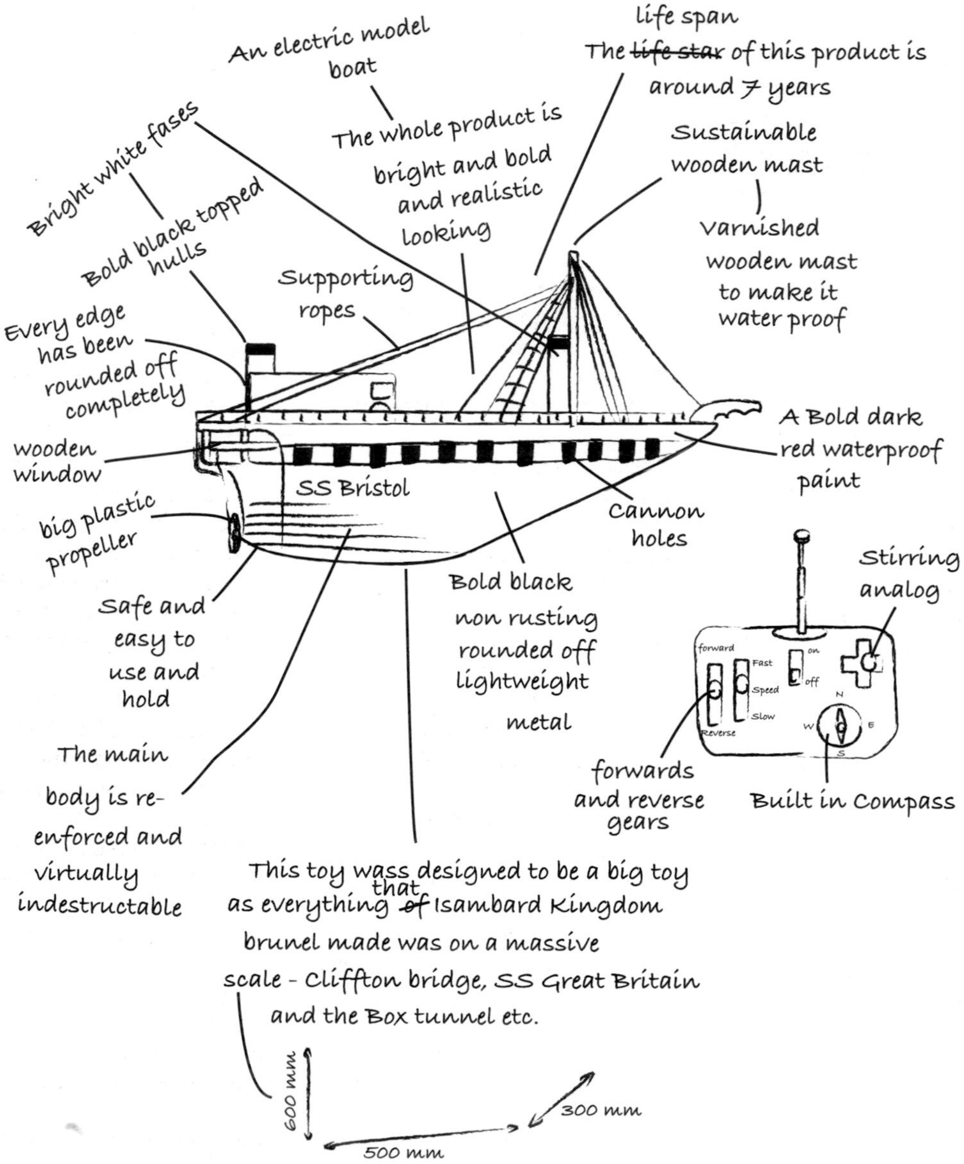

An electric model boat

Bright white fases

Bold black topped hulls

The whole product is bright and bold and realistic looking

life span
The ~~life star~~ of this product is around 7 years

Sustainable wooden mast

Varnished wooden mast to make it water proof

Supporting ropes

Every edge has been rounded off completely

wooden window

SS Bristol

A Bold dark red waterproof paint

big plastic propeller

Cannon holes

Safe and easy to use and hold

Bold black non rusting rounded off lightweight metal

Stirring analog

forward   Fast   on
Speed   off
Slow
Reverse

Built in Compass

The main body is re-enforced and virtually indestructable

forwards and reverse gears

This toy wass designed to be a big toy
that
as everything ~~of~~ Isambard Kingdom brunel made was on a massive scale - Cliffton bridge, SS Great Britain and the Box tunnel etc.

600 mm
500 mm
300 mm

Figure 11.8 Student response to Question 5(d)

INDEX

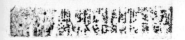